Fusion

FUSION

The search for endless energy

ROBIN HERMAN

Cambridge University Press

Cambridge

New York   Port Chester   Melbourne   Sydney

Published by the Press Syndicate of the University of Cambridge
The Pitt Building, Trumpington Street, Cambridge CB2 1RP
40 West 20th Street, New York, NY 10011, USA
10 Stamford Road, Oakleigh, Melbourne 3166, Australia

First published 1990

Printed in the United States of America

British Library Cataloguing in Publication Data

Herman, Robin
Fusion.
1. Plasmas. Nuclei. Fusion
I. Title
539.764

Library of Congress Cataloguing in Publication Data available

ISBN 0-521-38373-0 hardback

To Paul

Contents

Preface and Acknowledgments

Like many people, I came to fusion by chance and was intrigued by the utopian vision it offered of clean, endless energy from seawater.

I happened on it while living in Princeton, New Jersey. A friend's conversation piqued my interest in the nearby Princeton Plasma Physics Laboratory. I had never visited the lab and had no notion of the science pursued there. The only plasma I had ever heard of was blood plasma.

When I learned of the scientists' goal, saw their vast and complicated machinery, and began to appreciate the lifetime span of the work, I determined to somehow make a record of their quest. This book is the result.

I would like to emphasize that I took on the subject as an independent writer, an outsider with no particular point of view. I was not commissioned to write this book, nor do I have any affiliation or arrangement, formal or informal, with the institutions mentioned within its pages–although I did graduate from Princeton University, where major fusion research is under way. I hope this does not prejudice my account of its fusion project.

My inquiry took me on a journey back in time to speak with the founding fathers of fusion research. It also took me to fusion laboratories in England and Japan as well as the major U.S. labs on both coasts. I interviewed more than 150 scientists, engineers, government officials, and industrialists working on fusion in more than a dozen countries.

Because controlled fusion's history reaches back only to the 1950s, I was able to obtain recollections from participants in most of the science's major events. Thus this book is primarily a firsthand account. Nearly all quotations are drawn from personal interviews with the speakers. A few are taken from written sources that are either cited in footnotes or named in the text.

In addition, I am greatly indebted to several authors whose works on fusion preceded mine–Joan Lisa Bromberg who exhaustively details the American fusion program in *FUSION: Science, Politics, and the Invention of a New Energy Source,* a work commissioned by the U.S. Department of Energy; John Hendry, an historian for the U.K. Atomic Energy Authority, who discusses the world's earliest fusion research in *The Scientific Origins of Controlled Fusion Technology;* and T.A. Heppenheimer, whose *The Man-Made Sun* takes a look at fusion politics in the United States.

I wish to thank the fusion community in general for its courtesy, openness, and cooperation as I pursued my research. In particular, I must thank Harold Furth and Don Grove at the Princeton lab for their help and encouragement as well as Jane Holmquist and her industrious library staff at the Princeton lab without whom I could not have written this book. At the JET lab in England, I am grateful for the help offered by Paul-Henri Rebut, Roy Bickerton, and Bas Pease. At California's Livermore lab, I was graciously received by Dick Post, Dave Baldwin, Ken Fowler, and Erik Storm. In Japan, I express my deep appreciation to Yasuhiko Iso and Masaji Yoshikawa for showing me the Japanese fusion program.

For their hard work and faith in this project, I wish to thank my agents, Janet Pawson and Arthur Kaminsky, at Athletes and Artists in New York.

Thanks also go to my friend Lynne Yerby who helped me revise my first draft and encouraged me thereafter. And thanks to my friend Sharon Naeole who had the suggestion, five years ago, that I visit the Princeton Lab.

Most supportive in my work–and responsible for the fact that I completed it all–is my husband, Paul Horvitz, who was the chief editor of the manuscript. He worked long hours at no salary to make this book what I hoped it might be. He offered steadfast and loving support throughout the five years it took to complete. I thank him with all my heart.

Paris
January 1990

R.H.

Prologue

In 1984, on a crisp September morning, a fleet of tour buses set out from London bound for a rather ordinary village in the countryside near Oxford. Aboard were several hundred scientists from around the world who had worried over the years that this day might never come.

They were Frenchmen, Russians, Americans, Japanese, Israelis, and Italians. They looked out of the broad bus windows and exchanged pleasantries in serviceable English. Their true common language, however, was the vocabulary of a science that the world at large had little inkling of, an obscure language unknown and unspoken outside their tight fraternity.

Anyone stumbling on this fraternity would have discovered men who measure time and commitment in decades. They spend their days, and often their nights, inside huge laboratories housing a jumble of machinery that might one day bring about a new age of energy. Long ago they had reached across international borders to collaborate so that now, within their society, East-West political boundaries had all but vanished.

As the caravan of buses advanced along the highway, London's industrial fringes gave way to gentle green hills populated by lambs and woolly sheep, horses and pink pigs in the mud. Large cumulus clouds rose on the horizon, but at closer range, they proved to be great billows of steam erupting from the six giant cooling towers of the coal-fired Didcot power plant. The motor coaches passed another set of fields and hedgerows and then took a turn at a thatched cottage, pulling within sight of their destination: a blue lab building with gray and white tiers that loomed like a battleship over the fields.

On the lawn, an immense canvas tent had been raised for an outdoor banquet. There were tables covered in formal white cloths, crystal and china, and a lavish buffet of meats and cheeses. In front of the tent, chefs in white hats tended two roasted pigs turning on spits over open charcoal fires. A brass band was tuning up.

Outside the lab entranceway, flags of twelve nations flapped in the breeze. The scientists stepped off the buses and walked through the glass doorway of the lab, down the carpeted hallways past portraits of Queen Elizabeth and President François Mitterrand of France, to the mouth of a cold, cavernous hangar. Once inside, they looked up and stopped in their tracks.

Before them was the mammoth machine they had come to see.

Photographs had circulated, but nothing could convey the massiveness of the 3,000-ton device. It rose high and slender above the concrete floor, towering forty feet. The machine was a doughnut of steel caged by bright orange buttresses that pointed sharply toward the ceiling. The verticality of the structure directed all eyes upward as do the spires of a cathedral. The effect was uplifting, even majestic.

"Huge!" remarked Vladimir Muchovatov, a bespectacled Muscovite, his head tilted to take in the view. Phil Efthimion, a young American, started circling the thing. "Holy cow!" he exclaimed, "They said it was big . . ."

The machinery in total was even more extensive than what could be seen in the concrete test hangar. The full hardware systems protruded deeply into floors above and below and pushed out to adjacent rooms, eventually reaching high-voltage power lines linked to the laboratory's own substation, primed with electricity from nearby Didcot.

"Very impressive," the Siberian Dmitri Rjutov remarked. "A great thing for the community. It's not just a fantasy. After thirty years we have something impressive . . ."

The proud laboratory director, a portly German named Hans-Otto Wuster, had corralled Don Grove, a high-ranking American, sweeping him up on a personal tour. Paul-Henri Rebut, the Frenchman who had designed the machine and directed its construction, offered ebullient commentary to the congregation that remained.

The machine had been built by European scientists and engineers, but in a very real sense the achievement belonged to the entire gathering of international pioneers, to the fraternity that had convened in Great Britain from similar laboratories in Kharkov in the Soviet Ukraine, from Kyoto, from Boston, from Princeton and Livermore, from Haifa, Rome, and Beijing.

What they saw before them was an outgrowth of all their past work, the theories and the experiments, the failures – the many failures – and the successes. They circled the huge device in awe, then walked back outside to celebrate on roast pig.

The machine in the English countryside was called the Joint European Torus (JET), a billion-dollar experiment designed to show that fusion energy, the energy source of the sun and all the stars, could be created artificially on earth.

To the physicists who came that day, JET represented a kind of dual miracle. For decades their governments had allowed them to cooperate at a certain level, freely exchanging theories and data about the elusive science. But JET was more than cooperation. It was collaboration, a truly international project. Twelve European nations had signed the agreement to construct JET, contributing money and manpower to run it. That was the political miracle.

It was a scientific miracle, as well, that after more than thirty-five years of frustrating research, an experimental machine had been built with the potential, they believed, to meet the scientists' long-sought goal. The mission of the new machine was the same as the smaller test reactors in many of the industrial nations, forming a painfully slow evolution of the science of fusion. The mission was and remains this: to show that manmade fusion energy is scientifically possible and to deliver to the world realistic hope for an age of endless energy.

Fusion is the reaction not of atoms split but of atoms joined, atoms forced together in the massive furnace of the sun. The scientists' aim is to create minuscule fusion reactions inside a machine and to convert the prodigious energy released into an inexhaustible supply of electric power.

What's more, it would be relatively "clean" electric power. Fusion would produce none of the highly radioactive waste that the world's troubled fission plants churn out. Nor would fusion emit any pollutants like the choking carbon dioxide fumes that coal-fired plants pour into the atmosphere, warming it in a greenhouse effect.

There was, to be sure, a radioactive side to fusion in the use of tritium fuel, a radioactive form of hydrogen with a short half-life of twelve years. Eventually, the interior walls of any fusion reactor would become radioactive from the energetic activity inside. These walls would have to be replaced and disposed of. But in comparison with fission reactors, fusion appeared relatively benign.

Fusion was and is a utopian vision: clean, safe, perpetual energy for mankind derived from the most abundant of fuels – the hydrogen atoms in seawater. It would also be politically equitable energy that would, its proponents assert, make any nation as energy-rich as oil-drenched Kuwait.

Yet cracking nature's fusion laws is proving so complex and subtle that a lifetime of work might not be time enough to uncover them.

Through the years, breathless announcements of breakthroughs in fusion have frequently been followed by embarrassing retractions. In 1989, the University of Utah's sensationalized "cold fusion" experiments with palladium in a room-temperature solution promised an upheaval in the science but, in the end, seemed only to confirm fusion's stubbornly utopian nature.

The men who gathered outside London had been working on traditional "hot" fusion since the days of Truman, Stalin, and Churchill. Their founding fathers, men in the generation of Princeton astrophysicist Lyman Spitzer and Soviet atom researcher Andrei Sakharov, had, by the century's last decade, either retired or died or, in Sakharov's case, spent years in exile.

To talk of fusion, then, is to talk not of a moment in science but of generations.

Its story also offers insights into human nature, into the search for scientific knowledge. It offers a perspective on human endeavor from people trying to create something for the benefit and security of the species against the deadline of their own lifespans. The struggle to grasp fusion is the perpetual human struggle against perceived limits, intellectual and corporeal.

A creed of faith in fusion was expressed in surprisingly similar ways by the far-flung members of the community. People of entirely disparate cultures, separated by continents and oceans, held many attitudes in common. In that sense, the fusion fraternity was really very small.

One of the American fusion pioneers, Richard F. Post, put it this way: A fusion reactor is "of ultimate importance . . . not just another gadget." And he wrote that, "Once we have learned how to tap it, fusion can supply man's needs for energy for thousands of millenia – until, and even after the sun grows cold."[1]

Trying to create something "of ultimate importance," however, had extracted years from the fusion scientists' lives. What had begun as a quick sprint to glory in the heady post-Hiroshima days of atomic research had turned into a frustrating lifetime's journey that initially unfolded in secret labs in the United States, the Soviet Union, Great Britain, and elsewhere. In fusion's initial years, the scientists were confident that a reactor was just around the corner.

In 1956, while working at the University of California Radiation Laboratory at Livermore, Post wrote: "It is the firm belief of many of the physicists actively engaged in controlled fusion research in this country that all of the scientific and technological problems of controlled fusion will be mastered – perhaps in the next few years."[2]

No one had counted on fusion's difficulty. Eventually, in 1958, all research was declassified at an historic gathering of scientists in Geneva. Researchers from East and West laid bare all their secret work and hoped that a collective attack would speed the advent of the fusion age. As the task hardened into decades of frustration, there were surprisingly few quitters. The lure of the mission was strong. For many physicists, fusion gave meaning to their work and to their lives.

Reflecting the passage of time, the scientists sometimes painted themselves on a grand, historic tableau.

They had a taste for likening their lives in their antiseptic laboratories to the lives of legendary figures, men with a righteous, nearly impossible cause, doing deeds for the greater good and a stab at immortality. As they described their work, they conjured up images of the Crusaders in search of the Holy Grail and Prometheus daring to steal fire from the gods to give to man. They were building a cathedral, a pyramid, a manmade sun, they liked to say.

If anything was truly shared by the close-knit society of international fusion physicists, it was this self-image. Though they worked half a world away from one another, Enzo Lazzaro, an Italian physicist at JET, and Kamitaka Itoh, a Japanese physicist, both said they felt like part of the crew on Columbus' great journey across the uncharted ocean. Fusion, said Itoh, is another "terra incognita" to be discovered and shared by all nations.

Monumental imagery was popular as well, for the reality of fusion research was that it would take more than one generation of scientists to complete the task. Yasuhiko Iso, director of Japan's largest fusion laboratory, made the point with an illustration he pulled from his desk. He had superimposed a scale drawing of his lab's big test machine onto a drawing of the fifty-three-foot-tall Great Buddha of Nara. He found it significant that they were almost exactly the same size. Dale Meade, a Princeton physicist, chose the pyramid metaphor. "I'm only going to see the bottom layers of this thing," he said. "It's going to be completed by a younger generation."

The investigation of fusion reactions began in the 1920s. The first experimental fusion devices had been built in the 1950s. The latest machines had each taken almost a decade to build. Added to that were years more of experimentation before the next-generation test reactor could be conceived. Even the scientists' best estimates put commercial fusion power well into the twenty-first century.

In human history, few other technological projects had taken that long. The Great Pyramid of Giza was completed in just thirty years (albeit through the slave labor of 100,000 men) and despite popular conception, the grandest Gothic cathedrals did not require centuries to build. Most were finished within forty years.

Mastering fusion and putting it to use, the researchers asserted, was the greatest scientific problem ever tackled by man. Even by 1990, it was simply impossible to know whether the right political, economic, and scientific circumstances would ever converge to enable the building of a commercially viable fusion power plant.

Year in and year out, faith in their science kept the fusion researchers moving ahead. Day by day, however, they were driven by a more mundane struggle. No matter what sort of test machine they worked on, all were trying to accomplish step one in the kindling of a fusion reaction – controlling a superhot gaseous cloud of particles called *plasma*. The sun, physicists theorize, is nothing but plasma. Heat a confined cloud of plasma and a fusion reaction would result. It sounded so easy.

Yet plasma behaved with such violence and complexity inside the experimental reactors that nearly four decades of work had not been enough to decipher the physical laws that governed it. The study of this cloud came to be called *plasma physics,* a completely new, uncharted branch of science for the post-World War II era. Heating plasma to a temperature of hundreds of millions of degrees and holding it together in a dense cloud was a challenge that many of the physicists found both maddening and beguiling.

Phil Efthimion, the Princeton physicist who had greeted JET with an all-American "Holy cow!", was one of the plasma cultists. A second-generation fusion scientist, Efthimion maintained a childlike wonderment with the world and it showed in his large brown eyes and warm, upturned smile. He was open to experience.

What was so fascinating about plasma physics? Although we live in an environment of earth, water, and air, Efthimion's physics education had revealed to him that the rest of the universe was mostly composed of something else, something that he could touch only with his imagination. Whatever was out there in the night sky, the stars and the stardust between them, was plasma, clouds of particles that could not exist naturally on earth.

"The thing is, that once you start to think about it and appreciate it, you realize your whole world is dependent on a plasma," Efthimion said, referring to the sun. "Though it may be foreign to you now on this planet, your whole existence and the planet itself essentially has been predicated on the existence of this plasma; the creation of the world was predicated on a plasma state."

The sun's plasma existed under conditions so extreme compared to human experience as to be incomprehensible, temperatures of up to 14 million degrees centigrade where matter was unrecognizable as elements or even atoms. Everything was reduced to subatomic particles that moved thousands of miles in the blink of an eye, giving off energy when they fused that radiated to a planet 93 million miles away.

The study of giant plasmas was the realm of the astrophysicists, distant spectators to the gaseous clouds of stars and galaxies. Plasma existed on earth only, perhaps, in the momentary flash of lightning bolts, electric charges of such magnitude that the atmosphere through which they passed would be heated into a fleeting plasma.

With the advent of plasma physics, scientists like Efthimion learned to make plasmas in the laboratory, playing with stardust in their giant machines. There was, said Efthimion, "a kind of exhilaration" about being able to study phenomena for the first time and, even more so, to explain those phenomena.

It was irrefutable that uranium and fossil fuels would eventually "run out" – that is, become so difficult and expensive to mine as to be useless as energy sources. The people who make a business of studying global energy needs were willing to concede that much. The more difficult question was: at what point would the depletion become so critical that new energy schemes like fusion would be necessary and economically attractive? How much time remained to the scientists to solve the riddle of the plasma? Fifty years? One hundred years? Three hundred?

Although most experts of the mid-1980s did not forecast an imminent crisis, no truly definitive assessment could be made. All the studies required certain assumptions about future energy use. Those assumptions were necessarily based on differing philosophies about how great an energy cutback industrial societies would tolerate, how quickly new technologies could improve energy efficiency, how much people would be willing to pay for energy, and how soon the developing nations would require a larger share of the world's energy resources. Other studies factored in the less quantifiable "costs" of environmental pollution and political instability as energy prices grow higher.

Oil might last for another seventy-five years under the right circumstances, some experts said.[3] The world's estimated coal supply was good for perhaps another 200 years, although much of it was hard to mine and there was a price to pay: pollution.[4] Without a "breeder" reactor program, the high-grade uranium necessary for efficient use of nuclear fission plants might run out in about fifty years.[5]

Other projections were more dire. A University of New Hampshire study showed that domestic oil supplies in the United States will be effectively depleted by the year 2020, and that the world will be out of oil by 2040.[6] A survey by the Institute of Gas Technology reported that based on proved reserves of fossil fuels, no growth in demand and "neglecting all the problems of location, development, and distribution" the world's supply of fossil fuels would be depleted by the year 2086.[7] With a mere 2 percent annual growth rate, the depletion date would advance to 2028.

Taking these projections together, the world's experts seemed to indicate that a new energy source would not become imperative until the second half of the twenty-first century.

To the fusion researchers, that verdict meant the energy crisis was now. Fusion could offer relatively clean energy from cheap and limitless fuel, but the steps to a viable fusion reactor stretched across generations. An intense, committed effort would be necessary to meet that deadline.

Instead, government support, at least in the United States, had been half-hearted for years, and the American program suffered what fusion scientists described as debilitating budget cuts during the Reagan and early Bush presidencies. Gripped by a sense of urgency, the scientists were thwarted in their efforts by political and economic short-sightedness. Although they were optimists about fusion power, the plasma physicists came across as pessimists about the world's future energy picture. To politicians, they could not help but sound like prophets of doom.

Privately, many said that nothing would help as much as another Arab oil embargo, something to give the public a good scare, a taste of the inevitable. The day would surely come, be it 50 or 100 years away, that there really would be no oil at the pump. Fusion, they argued, had better be ready.

In 1984, the gathering of physicists at the JET celebration was really little more than an excursion from the larger purpose in London, the biennial meetings of the International Atomic Energy Agency. It was better known among some of the participants as the "Plasma Olympics."

The meetings drew the upper echelon of the world's fusion scientists, all armed with their latest data and theories. They were the geniuses of plasma theory, the hardheaded experimentalists, the modest university professors, and the boastful directors who ran the world's foremost fusion laboratories.

In modern fusion there were dozens of small labs around the world. Among the heavyweights, however, there were but four: the joint European program based near Oxford, the United States' big lab at Princeton, the Japanese Atomic Energy Research Institute north of Tokyo, and the Soviets' Kurchatov Institute of Atomic Energy in Moscow. Each had cost billions of dollars over the years to build and operate, running round-the-clock shifts under the eye of government bureaucrats. Each employed hundreds of scientists, engineers, and support groups. And each had come to the Plasma Olympics with teams of researchers and reams of laboratory results to share.

The United States sent a large delegation. From Princeton came the clever and impudent Harold Furth, director of the lab, and his deputies, Don Grove and Dale Meade, along with several rising young stars such as Phil Efthimion and Rob Goldston. The Lawrence Livermore National Laboratory in California sent fusion director Ken Fowler. His crew was experienced with a myriad of innovative fusion devices, including some using laser beams. From the University of Texas came theorist Marshall Rosenbluth, fondly nicknamed by the Russians "The Pope of Plasma Physics."

The Moscow Institute team, famous for its theorists, was headed by Yevgeny Velikov, a plasma physicist who doubled as science advisor to General Secretary Mikhail Gorbachev. The Japanese sent Masaji Yoshikawa, manager of what was then a huge, nearly completed test reactor. He and much of his staff had spent years working in American laboratories before building a powerful rival program.

And from Culham Laboratory, the venerable British institution that now shared its site with JET, there came lab director Sebastian Pease and a stable of experimentalists like Derek Robinson and Nic Peacock who had played crucial roles in fusion's history. JET, of course, sent nearly everybody, from Wuster and Rebut at the top to the experimentalists who ran the night shift.

For all the information that poured forth at the Plasma Olympics there were rarely any big surprises. Gossip spread quickly in the post-secrecy era. The scientists kept in constant touch by telephone, telex, facsimile machines, and frequent personal visits.

Fusion's ranks were replete with friendly double agents. It would be difficult to keep a secret, and no one really tried.

The culture of fusion was international and so too was the language, a kind of scientific Esperanto. To the uninitiated, the alien tongue seemed truly bizarre.

At Princeton's regular Monday morning meetings, for example, they talked about the "banana regime," "fishbones," "giant sawteeth," and the length of the "flattop." The word, *mode,* was big. There was the high mode, the low mode, the ballooning mode, and the tearing mode. Sometimes they would muddy the debate further by using abbreviations. There was "MHD instability," not to be confused with "DCLC instability." There was ECRH as opposed to ICRH.

And of course there was the dreaded "resistive kink instability."

To understand the language also required an appreciation for the otherworldy sense of proportion that came along with it. A 10 million degree plasma was considered "cool." A grain of metal that fell off the interior womb of the machine was a "boulder." A plasma that was floating somewhere between 24 and 25 centimeters into the chamber's center was as "lost" as a cloud over the Pacific Ocean.

It was apparent from the language of plasma physics that the field was still a new, evolving science. In a desperate effort to communicate with one another about new sights and ideas, the physicists appropriated whatever words and images came to mind. Once spoken at a meeting or written in a paper, the impromptu phrases became accepted terminology.

The etymology of the phenomenon called "Marfe" is a case in point. A peculiar light burst emitted by the plasma, Marfe was first spotted by Steve Wolfe in a Massachusetts Institute of Technology machine. Since only Wolfe's instruments detected it, however, his colleagues were doubtful and, he said, referred disparagingly to the effect as a "Wolfie." When Earl Marmar, another MIT researcher, saw it too, they called it a "Marmarism." Both scientists took offense at the jokes, so their supervisor devised a compromise and dubbed the effect a "Marfe." That play on the names seemed sufficient until Marmar and Wolfe submitted a scientific paper on the subject in which they described "Marfe" as an acronym for "multifaceted radiation from the edge." Presented in serious fashion, "Marfe" thus entered fusion's opaque lexicon.

Apart from the mysterious force that rattled around inside their machines, the world's fusion scientists recognized another common adversary: government.

Fusion research had begun in small laboratories with machines that scientists were able to build by hand. But after the first decade, the devices were multimillion-dollar projects, and for that kind of financing the scientists could find only one backer. During the 1980s, world fusion researchers were spending approximately $1.3 billion annually, nearly all of it government money. The research was in far too basic a stage for commercial industry to take an interest, except in Japan where industry representatives were among the initial planners. Occasionally oil companies and utilities sent their people to fusion meetings, just to keep tabs on the slow pace of research.

No vocal public constituency for fusion existed. By the 1980s, memories of the energy crisis had faded and the environmental movement seemed to have lost the confrontational verve of the 1960s. Only on occasion did the fusion scientists' mission inspire advocates outside its close fraternity. Even then, at least in the United States, it had been a strange melange of fringe characters who rose to advocate a fusion future, often with little or no encouragement from the fusion community.

Bob Guccione, the publisher of Penthouse magazine, sank $16 million into private fusion research, hoping to reap a fortune from an historic technological breakthrough. The presidential candidate, political extremist, and self-styled economist Lyndon LaRouche had a legion of followers who set up tables at airports and distributed leaflets touting fusion energy among his many causes. The photographer

Ansel Adams, famous for his black-and-white landscapes of the American West, was a diehard fusion advocate.

But government was the only natural source of continuing fusion support, even though its enthusiasm ebbed and flowed. Lyman Spitzer, the professor of astrophysics who conceived ground-breaking, secret experiments at Princeton in 1951, had to contend with overzealous bureaucrats who expected fusion immediately. After a subsequent period of government disinterest, a brash energy official named Robert Hirsch forced the Princeton scientists of the 1970s to build a bigger machine than they felt was justifiable. The physicists derided Hirsch as "a young man in a hurry."

"If you don't hurry, you're not going to get to the point where you find out whether you were right or wrong – so you can change course and not waste time," Hirsch said in an interview years later, looking back at the Princeton decision. The first, ardent researchers like Spitzer, Hirsch said, found out that fusion "was horribly complicated." As a result, Hirsch added, "they were embarrassed and forced to go back to basics."

"They developed an attitude and a culture that everything that you do should evolve from an understanding of the basics, as opposed to trying to jump right to the thing that will work. And so the program became a plasma physics program instead of a fusion research program." According to Hirsch, that could not be allowed to happen. Washington had to keep an eye on the physicists.

During the Carter, Reagan, and Bush administrations, government agencies and research labs kept close watch, as well, on fusion developments in other countries and were frequently influenced by events and decisions abroad.

Even though fusion support fluctuated over the years, the scientists never claimed that a lack of government funds was the reason they had not been able to develop a fusion reactor. Bureaucratic conflicts made development more difficult, perhaps even slower, but unlimited financing was not the answer.

The question of money was put to Harold Furth, the director of the Princeton lab. "If the Martians were attacking," he said in his typically droll manner, "if money were no object and the military wanted a working fusion reactor by the year 2000, there is no question we could have it. By the year 2000 we could build such a BIG turkey." The point is, he said, that government does not want "a hugely expensive thing that produces a modest amount of power. Money will solve A problem, not THE problem."

THE problem was not just to build a machine that could make fusion. A giant, energy-gorging fusion reactor that produced a trickle of power could be built overnight. What society needed and economics demanded was an efficient machine that could generate large amounts of fusion power in relation to the electric energy needed to heat and hold the fusion-producing plasma. Solving fusion would take brains, not brawn.

From the secrecy years of Spitzer and Sakharov to the gossip generation of Goldston and Efthimion, the fusion scientists had created something enduring, if not a fusion reactor itself. They had created a fusion fraternity, an international achievement of which they are proud.

The scientists remained dedicated to one another and to their mission through scientific and political challenges. They clung to a creed that fusion was essential to human survival and that it was their individual responsibility to deliver fusion's promise. When it became apparent that fusion would take several lifetimes to complete, the responsibility fell to the community as a whole, renewed by young recruits to the mission.

Marshall Rosenbluth had retained his confidence in fusion's ultimate success through his years as a young and brilliant theorist to his mature role as a mentor to the new generation. It was obvious to him that since the universe runs on fusion it is "only natural" for mankind eventually to use the same process to satisfy energy needs. He said he still had "utter faith" in fusion and "absolutely no doubt that it will be. Whether I will live to see it," he added, "looks pretty doubtful."

Six months after the cheerful JET celebration, Rosenbluth and his fusion colleagues were in a more somber mood. They had gathered on the other side of the ocean, at Princeton, for the annual meeting of the international advisory board for Princeton's big machine.

After the meeting, a smaller group of old friends convened at Furth's apartment for a dinner. Rosenbluth and Rebut were there, along with the Briton Roy Bickerton, the scientific deputy for JET. Another guest was John Sheffield, a native Briton who had been on Rebut's design team for JET but now ran the United States fusion program at Oak Ridge and had become an American citizen. All five men were in their fifties; they had known one another through most of fusion's history.

The evening consisted mostly of trading stories of other foreign trips, of women met in distant cities during Plasma Olympics, of teenage memories of World War II – when Furth fled with his parents from Austria, when Sheffield's parents sent him to the English countryside for safety – of years spent working at other laboratories, adjusting to new cultures, pursuing old problems. The sense of time passing in their lives was heavy.

When the group sat down to dinner, Furth stood at the head of the table and raised a toast. "May your children live to see fusion power," he said, his eyes on his friends. Rosenbluth and Rebut immediately amended the toast. "May our grandchildren live to see fusion power," they chimed together. This brought a cynical laugh from Sheffield who added simply, "May our children have electricity."

As believers in fusion, they toasted the future.

They might have toasted the past as well. Some thirty-four years earlier, in the same wintry month of March, one of the world's most charismatic leaders had proclaimed that his nation had built a working fusion reactor. Ironically, it was this false claim of success that first propelled scientists the world over into the epic race for fusion power.

I

The invention of Dr. Spitzer

On Saturday morning March 24, 1951, the Argentine dictator Juan Peron called a news conference at Casa Rosada, the presidential mansion. By his side was an obscure Austrian physicist named Ronald Richter.

Before a throng of reporters with microphones broadcasting Peron's words to the nation, he declared that Argentina had built an atomic energy plant on the lake island of Huemul. The pilot plant, he said, employed a revolutionary and superior form of nuclear energy. Instead of pursuing nuclear fission as other nations had, "the new Argentina," as Peron put it, had taken the risk of attacking a different form of atomic reaction, the kind that powers the sun. On February 16 past, tests had successfully produced "the controlled liberation of atomic energy" without using uranium fuel. The reaction had required "enormously high temperatures of millions of degrees."[1]

Moreover, Peron declared, in the course of experimental work his scientists had proven that foreign researchers were "enormously far from their goal" of making a thermonuclear, or fusion, bomb. Of course, he said, Argentina would be using its new atomic capabilities "solely for power plants and industrial use." He called the discovery "transcendental for the future life" of his people and "for that of the world," and he said it would bring Argentina "a greatness which today we cannot begin to imagine."

To the international scientific community, it was an astonishing claim: Peron was saying that Argentina had a working fusion power plant.

The triumphant declaration created headlines worldwide. The *New York Times* carried an article at the top of the front page: "PERON ANNOUNCES NEW WAY TO MAKE ATOM YIELD POWER," and under that: "Reports Argentina Has Devised Thermonuclear Reaction That Does Not Use Uranium. Method Linked to the Sun's."

There were also comments, however, from several eminent American and European physicists expressing skepticism about the Argentine claim. Among the doubters was the Nobel Prize winner Enrico Fermi, who had helped invent the atomic fission bomb in the United States. Peron's technical man, Richter, had given no scientific details. The body of knowledge about thermonuclear fusion reactions was too limited for such an enormous technological leap to be credible.

Some atom researchers suggested that Peron's announcement was merely a political gambit to enhance Argentina's image as an independent world power. Whatever his intentions, Peron had not announced the creation of a bomb. That would have been an astounding turn of events, for the atomic club of the day included only two members, the United States and the Soviet Union. In secret laboratories, American and Russian scientists were intensely studying plans for more sophisticated bombs from fused atoms, but neither nation had yet produced a high-temperature "fusion," or thermonuclear, reaction – explosive or otherwise.

Lyman Spitzer, the thirty-six-year-old head of Princeton University's astronomy department, was one of those American scientists. The day after Peron's announcement, Spitzer was preparing for a weeklong ski trip with his wife to the quiet Colorado town of Aspen. As he was about to leave, his father telephoned from New York. Years later, Spitzer had a vivid memory of the call.

"He said, 'Well, I see the Argentines have beaten you to it.'

"And I said, 'What? Argentina?'

"And he said, 'Look at the *New York Times*.'"

As Spitzer pondered the news, he took the short mental step from bomb theory to ruminations about the peaceful uses of fusion power. Not long after Peron's boast, a full-scale program to build a genuine fusion energy reactor in the United States was quietly born.

Princeton University's secret fusion project, the first Stellarator, 1952. *Courtesy Princeton University Plasma Physics Laboratory (PPL).*

"It just looked like a kid made it. Very unimpressive. I didn't realize, you know, what it was going to come to."—JOE CSENTERI, electrical technician.

Spitzer's expertise was in a field of scientific inquiry called theoretical astrophysics. He speculated about the origin of the solar system and the explosive collisions of galaxies. His special interest was in the behavior of superhot interstellar gases like those of the sun and other stars. Spitzer wanted to understand how stars were formed.

Such esoteric study required a mind of extraordinary imagination and creativity. Spitzer felt that his greatest strength – and also his weakness – was his ability to imagine physical images of astral events. He could follow in his mind the development of a complex physical process and then translate that vision into concrete theoretical equations.

Astrophysics happened to be useful, as well, in examining manmade nuclear explosions, and Spitzer had agreed to lay aside most of his academic work for awhile to join the United States' crash program to develop a new, more powerful atomic bomb. The Russians had already

detonated their first A-bomb, and the United States was eager to be ahead in the arms race. The Americans wanted a hydrogen bomb, a thermonuclear device.

Spitzer and John Wheeler, a Princeton physicist who had recruited Spitzer for the bomb work, planned to set up a secret project at the university to study hydrogen-bomb physics. Their work would complement the efforts being made on the New Mexico desert plateau at Los Alamos under the direction of Edward Teller, the brilliant Hungarian-born physicist who had helped design America's first atomic weapon, the fission bomb.

The Princeton professors called their program the Matterhorn Project. To Spitzer, a serious mountain climber, it seemed as though designing the so-called "Super" hydrogen bomb would be a difficult and challenging climb.

When Spitzer got hold of the *New York Times* that Sunday after his father's phone call, he saw that Argentina was not claiming it had a hydrogen bomb. Rather, the article vaguely described a power plant driven by energy derived from the heart of the hydrogen atom, its nucleus. It was the same kind of nuclear fusion reaction that fueled the sun – the same reaction that Spitzer and Wheeler wanted to understand and translate into a hydrogen bomb.

As he headed for Aspen, Spitzer became engrossed in Argentina's momentous claim.

The resort was a quiet, simple refuge then, not the commercial ski capital it would become. As such, it provided the wiry, energetic Spitzer with a special atmosphere for thought. Whenever Spitzer took the ski lift up the mountainside, he found himself in enforced solitude. The chairlifts were designed to carry only one person, and it was a half-hour's ride up the mountain for every ski run down the slope.

Gliding uphill, looking out over the snowy landscape, Spitzer thought about thermonuclear power. If one could release and control the energy inherent in the hydrogen nucleus, it would mean that man had discovered an infinite power source, a machine that ran on the elements of water. In producing electricity, at least, such a machine would eclipse forever the need for coal, for oil, for uranium, for all the earth's depletable resources. The implications were profound and historic.

Like other physicists, Spitzer strongly doubted that Peron's scientists had unlocked the power of hydrogen. But if one were to design a machine that produced endless thermonuclear energy, how would one go about it? He had the perfect background for approaching this challenge. As an expert in the behavior of hot gases, Spitzer already knew the decade-old postulate by Hans A. Bethe explaining how the sun produces energy from hydrogen. It involved a rather straightforward understanding of the behavior of atoms.

The core of atoms, their nuclei, naturally repel one another. But inside the sun, lightweight hydrogen nuclei are so hot and moving at such fantastic speeds that when two collide, they fuse. This union forms a heavier nucleus of helium. The result is an instantaneous release of energy, reducing the combined mass of the new helium nucleus. The energy is given off as heat, visible light, and other radiations. Also thrown off from the reaction is a new, lighter particle – a fast-moving neutron. As Einstein had said, mass can be converted to energy. It was this fusion reaction of the sun that Argentina claimed to have mimicked.

Spitzer would have to learn how it might be accomplished on earth, and he began by calculating the tremendous heat needed for fusion. Using hydrogen, the least repellent of the elements, Spitzer found that under earthly conditions it would take a temperature of about 100 million degrees centigrade to make continuous fusion. That kind of heat was almost unimaginable. No container known to man could hold gases so hot. Any material on earth would melt, evaporate, disappear at much less than 100 million degrees. Besides, as soon as the hot hydrogen particles touched the cooler insides of any container, they would lose their heat, and the fusion reaction would stop immediately.

It was a question Zen-like in its simplicity. How do you hold a substance that cannot be touched?

As Spitzer considered this question, he recalled what he had read in a new book by Hannes Alfven, a Swedish scientist. Alfven had written about the influence of magnetism on hot gases in the cosmos. A superheated gas would become electrically charged, or "ionized." Charged particles could be attracted to magnetic fields. On one of his swings up the mountainside, as he thought about ionized gases, magnetic fields and Hannes Alfven, Spitzer found an answer: If superheated hydrogen gas could not be allowed to touch the walls of a container, Spitzer would simply make a container that had no walls. It would be an invisible "bottle" formed by magnetic fields inside a vacuum chamber.

Of course, it was just an idealized concept, this magnetic bottle. But it might work. In Spitzer's first visualization, the invisible bottle was a cylinder of infinite length with magnetic field lines running lengthwise within it. The hydrogen gas inside would be heated, and the charged particles should then cleave to the lines of the magnetic force and not stray beyond. As they raced around inside this "bottle," tightly confined by magnetism, the hot hydrogen nuclei would collide and fuse. Puff in tiny shots of hydrogen gas and the reaction would continue. Stop the fuel or turn off the magnets and the reaction would cool instantly and collapse.

If one made it to that stage, the real electricity for the nation's power grid would be easy. One of the byproducts of the fusion reaction is a fast-escaping neutron, a particle without any electrical charge. Neutrons would thus surge beyond the bounds of the magnetic bottle. They would speed out, crash into a neutron-absorbing liquid barrier, heat the barrier enough to boil water and create steam. The steam would drive turbines to produce electricity.

Lyman Spitzer suddenly lost enthusiasm for the Matterhorn weapons project. He had found a more interesting question. Instead of bombs, he now wanted to make a thermonuclear, or fusion, reactor – to use tiny amounts of fuel and contain the resulting energy rather than release it in a huge, split-second explosion.

"When we got back to Princeton I continued thinking about it, at night, when I couldn't get to sleep," he said.

Spitzer soon worked out how one might close the infinite "bottle" at its ends to form what he called "a finite system." Take the cylinder, he thought, and just join its two ends to form a closed, racetrack-shaped tube. Spitzer knew that this solution would require a complicated grid of magnetic fields to compensate for the bends in the bottle, just as a speedway banks its sharp turns. That was the cost of transforming an imaginary scheme into reality.

Within a month of his vision on the ski lift, Spitzer presented the newly formed Atomic Energy Commission (AEC) in Washington with a proposal for a twelve-foot-long tubular fusion device wrapped on the outside with electromagnets in the spiral-striped design of a candycane. Through the force of his enthusiasm and confidence, he quickly convinced the weapons authorities to include work on controlled fusion in the Matterhorn Project, at least for the next year. The weapons application of a fusion machine was unknown. Its economic and political impact would surely be great. As the civilian agency in charge of developing both nuclear weapons and peaceful atomic energy, the AEC saw a dual utility in controlled nuclear fusion.

Wheeler would handle the secret bomb theory work. Spitzer would concentrate on a fusion machine.

Seeing no military usefulness himself in controlled fusion techniques, Spitzer pressed the AEC to release his portion of Matterhorn from the bonds of secrecy. How could he recruit good people to join his team if he could not tell them what they would be doing? How could his small staff be expected by themselves to explore and apply an entirely new realm of physics? Pioneering science and secrecy were incompatible, he argued.

It was precisely this frontier aspect of the new fusion physics that worried the government. Who knew where it might lead? Perhaps the abundant and cheap neutrons produced by the reaction could be used to generate radioactive bomb-grade materials out of low-grade uranium.

On December 21, 1951, Thomas H. Johnson, director of the AEC's research division, wrote back to Spitzer, "The release of the fundamental idea and fundamental research information related to your project might weaken the future security of this country."[2] The answer to Spitzer's request for declassification was a firm no.

Secret research required out-of-the-way, off-limits places, so Spitzer located a university building about five miles from the main Princeton campus. It had formerly been a rabbit hutch used for animal research by the Rockefeller Foundation. Renovations were begun. The windows of the round metal hutch were sealed and blackened. The interior walls were striated with sensitive wire that would set off an alarm if breeched by intruders. A high metal fence topped with barbed wire was erected. Security clearances and badges were issued to Spitzer's newly assembled staff. Guards were posted at the entrance, and the work began.

Spitzer embarked on his project with the supreme confidence characteristic of government scientists of the day. Fusion looked to be solvable, and the bureaucrats in Washington agreed. Juan Peron had merely been duped into a premature announcement. In a matter of months after Peron's declaration alongside Richter, the Argentine press reported that the Austrian physicist had been jailed for misleading the president.

In Washington, timetables were established and budgets fashioned. Spitzer's plan called for three consecutive experimental fusion machines that would result in a prototype reactor in less than a decade. The scientific proof of controlled fusion's feasibility would come even sooner, he believed. Tom Johnson, the AEC research director, reported to the commissioners that it would cost about $1 million over the next three and a half to four years to answer the basic questions: Can a hydrogen gas at high temperature be confined in a magnetic field? If not, why not?

In 1951, there was no reason to doubt the inevitability of harnessed fusion power on earth. What Spitzer badly miscalculated, however, was the time it would take to bring about the new age.

Joe Csenteri could hardly wait to be allowed into the rabbit hutch. He had been hired as an electrical technician for a secret project at Princeton, and until he received his security clearance he could not be told exactly what he would be working on.

When he was finally ushered into the experiment room and saw America's first fusion research machine he felt puzzled and let down.

"You're wondering what it was," he said, "and then when you go in there – a two-by-four rack with some pipes, these receptacle switches. Nothing. It just looked like a kid made it. Very unimpressive. I didn't realize, you know, what it was going to come to."

There was nothing elegant or futuristic about the look of this secret, ground-breaking fusion device handbuilt by Lyman Spitzer and his staff. It consisted of a glass pipe, a vacuum system to pump the air out of it, a heating source and an electromagnet, all powered by an ordinary 220 volt socket in the wall. The glass pipe was two inches wide and about twelve feet in total length, bent in a racetrack shape. It was suspended by hooks and clothesline wire over a wooden table. The heating and magnetic forces came from wires and copper coils wound by hand around the glass. A vacuum tube poked up through a hole cut in the table. The pump was on a shelf underneath.

Despite its modesty, Spitzer's fusion device was designed to reproduce on a miniature scale the reaction that takes place in the sun and the stars. As a result, Spitzer gave it a grand name: the Stellarator.

Other new recruits found the science impressive, if not the machinery. George Martin, a technician with a physics and chemistry background, was excited by the advantages of fusion energy over fission power plants.

"It was like a giant jackpot," Martin said. "You can unleash a fantastic amount of energy, and it had what you'd call a lot of superior economic possibilities to fission. See, everybody's got water, but how many people have got uranium? There would be no 'have-not' nations, in a manner of speaking. It was such a big source of energy. Greed. We were overpowered by the riches." It was as though, he said, "a giant pool of oil" were under the scientists' feet.

"You think, 'Oh my! I've got to dig down to it.' We knew there was a big reservoir, and we just clawed frantically to achieve some progress."

The little fusion community inside the rabbit hutch worked closely and in isolation. There were just twenty or so men, the physicists recruited from the fields of cosmic science and nuclear weapons work, the technicians from local industrial plants. They held their meetings around a single table, "like King Arthur and the knights of the round table," said Martin.

Spitzer was unquestionably their leader, held in the highest regard for his intellectual vision and his unfailingly gentlemanly manner. Spitzer had a hand in every aspect of the experimentation. He unloaded equipment from truckbeds; he helped wind the magnetic coiling; he had suggestions for cleaning the vacuum tubing; and he coached the theorists in devising the physical principles that underpinned all their work.

"All of us agreed that Lyman Spitzer was really seven people," said Tom Stix, a cosmic-ray theorist who had planned to do bomb work but was sidetracked for life by a job offer from Spitzer. "There was the one who was head of the astronomy department, the one who was head of the lab, the mountain climber, the one who edited the Yale Daily News. He played piano. He sang Gilbert and Sullivan. And he turned out to be a marvelous administrator."

Spitzer's strongest interest appeared to be in fusion theory. While most scientists would say they could not believe a theory unless it was backed up by experimentation, Spitzer would joke that he never believed an experiment unless there was a theory to support it.

When Spitzer began examining the sun's hot plasma with the idea of applying its principles to an energy machine, his inquiries in the rabbit hutch opened a new branch of science. At the time, there were no textbooks describing the behavior of manmade plasma – after all, the subject was classified secret – and no formal theories about how plasma reacts to gravity, magnetism, and other forces. The world *plasma* itself was not yet in general use.

The term had originated in 1922 with the American physical chemist, Irving Langmuir, who was working at General Electric Research Laboratory in Schenectady, New York. Langmuir had been passing electrical discharges through gases and needed a term for the partially ionized gas that resulted. The way the gas contained different particles such as high velocity electrons, molecules, and impurities "reminded him of the way blood plasma carries around red and white corpuscles and germs," wrote a colleague, Harold M. Mott-Smith.[3] So Langmuir proposed the term *plasma*.

"But then we were in for it," wrote Mott-Smith. "For a long time we were pestered by requests from medical journals for reprints of our articles. The scientific world of physics and chemistry looked askance at this uncouth word and were slow to accept it in their vocabulary."

Instead, scientists referred to "hot ionized gases" in the select circles permitted by the government to discuss the subject. Spitzer was unaware, for example, that Edward Teller and Enrico Fermi and others had speculated about controlled nuclear fusion reactors at Los Alamos during the war.[4]

The rabbit hutch team was venturing into fresh intellectual territory. "It was a small group and we knew we were pioneers in what was going to be a big area," Stix later recalled. "It was marvelous that so little had been done. A small amount of work, relatively speaking, could have enormous impact, not only in getting things published but in patents as well. To get such rich rewards for your effort was great."

The open territory of fusion physics would eventually attract fellow astrophysicists, nuclear physicists, rocketry specialists, and others whose knowledge and curiosity brought them to its borders. It was a very small fraternity whose science and revelations would come to be known as plasma physics.

The initial optimism inside Spitzer's rabbit hutch was a product, in part, of the fantastic earlier strides made in atomic research. A decade before, the scientists who had set out to split the atom had made a quick success of it. Two and a half years after Enrico Fermi put together an experimental atomic pile in a squash court under Stagg Field at the University of Chicago, a pair of atomic fission bombs were dropped on Hiroshima and Nagasaki. The bombs provided dramatic, devastating proof that man had mastered the science of fission energy.

Radioactive uranium was a most cooperative fission fuel. Its heavy nucleus, loaded with extra particles, needed just the push of an additional neutron to become unstable and split part, releasing tremendous energy plus excess neutrons. These neutrons would in turn trigger more fission reactions in the fuel, causing a chain reaction. The energy hidden in the uranium nucleus was so great that just one gram of fuel was enough to destroy the entire city of Hiroshima.[5]

But the scientists' mastery of fission extended beyond its explosive uses. Even before the bombs were built, scientists had learned how to create and control fission on a small scale. It was a simple proposition. If excess neutrons were removed, the chain reaction would slow down. Fermi knew this from the start. By lowering a control rod of neutron-absorbing material into his atomic pile, he had been able to moderate and then stop the reaction.

This method was so technologically straightforward that it quickly encouraged the development of fission power plants. Immediately after Fermi's experiment, the U.S. government felt enough confidence in the new fission science to begin building fission reactors for military use. The chain reaction produced a byproduct, plutonium, that could be used as bomb material. The Hanford Engineering Works at Pasco, Washington, started operations in 1944.

Fission was quick. The government expected no less from fusion, and the fusion bomb experts complied.

At the Los Alamos national weapons laboratory, Edward Teller and his researchers were under no constraint to control their fusion reaction, as was Spitzer. They could employ large amounts of hydrogen and explode it in the atmosphere. To do so, Teller's team devised an ingenious way to recreate the tremendous forces and temperatures of the sun. They used nothing less than a nuclear fission bomb as a match to detonate a core of hydrogen fuel. The terrific force of the fission blast would instantly heat the core to fusion temperature while also compressing the hydrogen particles until they fused. If one used enough fuel – several grams of hydrogen was enough – the result would be a fusion reaction touching off a second giant blast.

On November 1, 1952, on the Eniwetok island chain in the Pacific, America's first hydrogen bomb was secretly tested. The explosion obliterated a small island and dug a crater a mile wide into the Pacific Ocean floor. Manmade, uncontrolled fusion was a powerful force indeed.

"No sooner was it done," Teller later wrote[6], "than every politician and every bureaucrat descended upon us saying, 'Now you must solve the problem of controlled fusion.'"

A theorist at heart, Spitzer nevertheless wasted no time attacking the practical objective of controlled fusion. His plan unfolded in steps, and the first was to create a plasma inside the tabletop Stellarator.

Although the sun was his theoretical prototype, it was impossible to copy in practice. It was a matter of scale. The sun's fusion furnace is vast and prodigious. It converts mass into energy at the stunning rate of 5 million tons per second. The sun is so big – its total mass is 300,000 times that of earth – that it creates an overwhelming gravitational force. This force holds the roiling mass of plasma together at ten times the density of lead, guaranteeing abundant collisions of particles. The great density permits the sun to percolate along at a "cool" 10 million degrees or so. Even if the particles could escape, where would they go? The sun is hundreds of thousands of miles across, so large that fleeing particles are confined long enough for collisions with other fleeing particles to be virtually assured.

High temperature. High density. Long confinement time. These are the three pillars of a successful fusion reaction, and Spitzer had to find some way to meet these demands inside a machine. The fusion demonstrated by the sun was explosive, not controlled. For bomb builders like Teller, a fission blast could easily create the terrific heat, density, and confinement needed to ignite a core of hydrogen. For Spitzer, the goal was a small, steady, absolutely controlled fusion reaction, a reaction that could be started and stopped by the turn of a gas jet, the flick of an electrical switch.

That was part of fusion's beauty. Ultimately, the natural tendency of the atomic nuclei to repel one another made the fusion reaction inherently safer than fission. If an accident allowed the hot gas to escape its container, the reaction would cool instantly and come to a dead stop. There would be no runaway chain reaction, no possibility of a "meltdown" of atomic fuel. The potential for a safer kind of nuclear reactor made fusion all the more compelling.

At first, Spitzer simply wanted to show that his invention could create and hold a hot plasma "at least a few seconds" at 1 million degrees, as he had told the AEC. So few particles of hydrogen were used that relatively little power was needed to raise them to very high temperatures. Initially, Spitzer used radio frequency waves that would transfer their energy to the gas. The same principle would later be used to heat food inside microwave ovens. The tiny particles of hydrogen had to be heated inside a pristine vacuum because any interloping elements would steal heat from the reaction. Ridding the vacuum of these "impurities" proved to be a tremendous engineering challenge that would plague fusion for decades to come.

Perhaps most frustrating of all was the challenge of observing and measuring what was going on inside the Stellarator without disturbing the gas and ruining the reaction. To know that he had even made plasma, Spitzer would have to rely on a crude tipoff, the light coming from portholes in the machine. Just as neon emits light energy when electricity heats it to a "cool" plasma state inside a bulb, so does a hot hydrogen plasma give off light. A very bright light in the Stellarator followed by a gradual dimming would mean that gas was being converted into plasma.

The Stellarator – "Model A," they called it – was ready for experiments in the fall of 1952. Inside the black-windowed rabbit hutch, the room was darkened. Then, according to Spitzer's log book, a "light purple glow" began to appear in the Stellarator's portals. The glow grew fainter, then disappeared abruptly. Sometimes it was brilliant, sometimes dim, but it was clear that the Stellarator could easily make plasmas, however fleeting.

Eventually the Stellarator produced plasma of perhaps half a million degrees centigrade, the scientists estimated. But the hot gases disappeared almost instantly, lasting less than a thousandth of a second before escaping the grip of the magnetic fields and hitting the cool walls of the vacuum vessel, quenching the reaction. This was not the way to make fusion. Spitzer's physicists were puzzled. Something was happening to the plasma for which their theories could not account. Still, there was enough promise in the invention to warrant building a larger machine, one that Spitzer hoped would solve the problem.

The Model B Stellarator was a souped-up version of Model A – stronger magnets, better heating, and a cleaner vacuum with fewer impurities. The heating trick was a simple pulse of electricity shot directly into the plasma, just as a current warms the metal coils inside a toaster. The vacuum engineers decided to vaporize impurities with high heat – much as self-cleaning ovens would later work – and suck out the vapors with a pump.

Model B began operations in 1954, and it reached the 1 million degree plasma for which Spitzer had aimed. But it was no triumph, for the invisible magnetic trap still held the plasma for less than a thousandth of a second. The scientists had made plasma, but there was no time for a fusion reaction to take place.

In theory, the charged particles of the plasma were supposed to move on a steady path. Instead, the plasma as a whole was in an uproar. In Spitzer's words, it developed "shimmies and wiggles and turbulences like the flow of water past a ship." There was "no obvious reason" why it had to be that way, he said, but it was ruining the reaction. The particles were escaping the magnetic force lines and striking the interior walls of the Stellarator, making further heating attempts impossible.

The nascent science of plasma physics had treated plasma as a collection of independent particles, but it was suddenly inadequate to explain the wild, turbulent fluid inside the Stellarator.

In those early years, fusion was fast evolving into a tougher scientific problem than anyone had predicted. Theories proved unworkable. New theories sprang up. At the same time that experimental fusion machines were being developed at Princeton, other small, secret programs in controlled fusion were getting under way at a handful of government labs such as Livermore, in northern California; Los Alamos, New Mexico; and Oak Ridge, Tennessee. Scientists there were experimenting with magnetic fields patterned differently from the Stellarator. Ideas to explain and control the behavior of plasma floated through a tiny cadre of youthful American researchers.

Like a concerned uncle, Edward Teller had kept abreast of the controlled fusion work, and he had doubts about the new experimental machines. In 1954, at a meeting of the fusion community held at the Princeton Gun Club, a wooden lodge on a field near Spitzer's rabbit hutch, Teller enunciated his theory of chronic "instabilities." Trying to confine plasma with magnetic field lines, he told the scientists, was like trying to hold a blob of jelly with rubberbands. More understanding was necessary to find a technology that would work.

Teller's new theory was troubling, and it took some typically futuristic brainstorming by Lyman Spitzer to provide an answer. Soon after the Gun Club meeting, Spitzer and three of his staffers began what might have been a dreary, three-day train trip out to the Livermore lab in California. Instead, the journey developed into a rolling, marathon think tank. Spitzer liked to stretch the limits of ideas, and one of his common techniques was to remove mundane assumptions that barred the way to creative thinking. Even before the age of space flight, for example, he had proposed that a telescope be launched into orbit to view the heavens without the interference of the earth's atmosphere.[7]

On the train, Spitzer visualized an imaginary power reactor that produced unlimited electricity without cost. Could one take that free power and make it easier to manufacture a new liquid fuel for mobile units like cars? On another subject, the team discussed micromanufacturing – machines that could replicate themselves, each on a successively smaller scale.

"You see how far you could push that," Spitzer explained, "and it ends up, for example, you make little airplanes the size of flies. Maybe you could use them for chasing insects."

In this rich, speculative atmosphere the subject of Teller's instability theory came up and by journey's end Spitzer had devised a scheme to redo the magnetic fields of the Stellarator. He would add a second set of electromagnets in a different configuration to provide tighter field lines. When he returned to Princeton, his team of theorists applied the new magnetic pattern to their calculations and drew some optimistic conclusions. Spitzer felt the Teller challenge had been resolved.

It was a small step forward. Even with the problems of the Model B Stellarator, the AEC agreed to allow design work to proceed on a Model C, projected to cost more than $20 million. As Spitzer had originally planned it, the Model C would be a prototype fusion reactor. Instead it was now crafted as another research machine, but so large that it would be capable of producing plasma of about 100 million degrees centigrade – his original temperature estimate for a continuous fusion reaction. A new building was planned for the Model C, with its own small power station.

After five years of secret work inside the rabbit hutch, Spitzer's confidence in the ultimate success of the Stellarator was unabated. But the Princeton researchers had not yet answered the crucial questions "Can a hydrogen gas at high temperature be confined in a magnetic field? If not, why not?" They were already falling seriously behind the original timetable.

To Teller and Spitzer, this was a mighty argument for opening the locked doors of fusion research, for bringing fresh ideas and broader exchanges to the field, for declassifying the science of plasma physics. Both were on a committee created by the AEC to study the possibilities. Looking into the future, Spitzer saw secrecy not as a layer of protection but as a real barrier to understanding.

"I remember," he said, "we recommended taking steps toward declassification just for the reason that there seemed to be basic problems in the way of achieving controlled thermonuclear fusion." Cracking the code of plasma over the long haul, he felt, required "a free exchange of information" on the widest terms possible.

"The rewards of secrecy didn't seem to be that great, and the penalty in slowing down progress might be quite severe."

But the powers on the Atomic Energy Commission were too leery of allowing fusion's promise to pass into Communist hands. The closed fusion community would have to solve its own problems.

2

Behind closed doors

Lyman Spitzer and Edward Teller were not alone in their desire to see fusion released from the fenced laboratories into the open field of international exchange. There were other men in distant, undisclosed places who studied the intractable plasma.

One was Igor Kurchatov, a tall Russian physicist with a long flowing beard who headed the Soviet Union's nuclear weapons program. It was a secret program, of course, carried out at the "Laboratory of Measuring Devices." But in the spring of 1956, Kurchatov ventured out. He accompanied the Communist Party leader Nikita Khrushchev and Soviet Premier Nikolai Bulganin on a trip to Great Britain where he arranged an invitation to speak at the Harwell Atomic Energy Research Establishment west of London. There, in the very citadel of Britain's secret nuclear research, the Russian physicist stunned his audience by delivering a lecture entitled "On the Possibilities of Producing Thermonuclear Reactions in a Gas Discharge."[1]

In one bold stroke the Soviet Union overturned the rules of the Cold War information game. Kurchatov laid bare the fact that for the past six years his nation had been conducting research in controlled fusion.

With breathtaking candor, Kurchatov described one set of experiments in some detail. Then he called upon scientists the world over to remove the security curtain from controlled thermonuclear research and embrace cooperation on this peaceful pursuit.

"He created quite a stir for awhile there," recalled Alan Ware, a British researcher who had been working on secret fusion schemes himself. "We

Lyman Spitzer, Jr., inventor of the Stellarator and founder of Princeton University's fusion program, code-named "Matterhorn." Courtesy PPPL.

"When you're climbing a mountain, until you're out on the summit, you're not really certain you're going to make it."—LYMAN SPITZER.

were all briefed ahead of time, before he gave his lecture, that we could listen intently, but we weren't to say anything about what we were doing."

The speech was a dramatic gesture in a dramatic setting, and it was intended to goad the West into declassifying its secrets behind the lead of a peace-loving Soviet nation. Moreover, it was a direct appeal to this narrow body of scientists to form a world community with a common goal. When copies of Kurchatov's speech made their way to the United States, Spitzer, Teller, and other American scientists reported to the Atomic Energy Commission that the Russian accomplishments in basic fusion appeared genuine and impressive.

The careful secrecy practiced by both the United States and Great Britain could not prevent the inception and growth of an idea whose time had come. Controlled fusion, it turned out, was not a wild scheme that occurred to one man on a ski lift. It was a logical outgrowth of the new nuclear science. The United States had been well aware that its ally, Great Britain, had been working to design a fusion reactor. What more had the Russians learned about the elusive plasma, the Americans wanted to know. How close were they to a real reactor? If those answers were not immediately available from Kurchatov, at least it was now certain that the Soviet Union was in the race, too – that the seeds of fusion power had blossomed almost simultaneously in the East and the West.

In his day, Igor Kurchatov was the most influential scientist in the Soviet Union. But his speech at Harwell, delivered in Russian with accompanying written translation, had identified two other men as the founders of Soviet research in controlled fusion. They were Senior Academician Igor Tamm and his young protege – a new name to Western scientists, Andrei Sakharov. As a human rights activist in later years, Sakharov would win the Nobel Peace Prize but land in the cold internal exile of Gorky as a symbol to the West of Communist repression. In 1950, however, he was a respected scientist of exceptional promise working at the fenced-in Moscow weapons laboratory headed by Kurchatov.

Sakharov was deeply involved in the development of the Soviet hydrogen bomb, for which he later thrice received the title of Hero of Socialist Labor, his country's highest civilian honor. Sakharov and Tamm, however, took an interesting sidetrip into controlled nuclear fusion. Like Spitzer, they speculated that magnetic fields could trap hot hydrogen plasma. They laid out the fundamental physics, and then Sakharov went beyond idealized laws to design an actual power-generating fusion machine.[2]

He showed these ideas to Kurchatov, a vigorous administrator who was nicknamed "the Beard" by his staff. The outcome was recorded by Kurchatov's deputy and official biographer, I.N. Golovin.[3]

It was New Year's Eve in 1950. Kurchatov lingered in the Moscow laboratory, contemplating past achievements and musing over the lab's future as he stroked his beard, a sign of contentment, Golovin noted. He asked his deputy what the theorists thought of Sakharov's new idea. Golovin replied that Kurchatov had already solved the first big problem of the nuclear age, designing an atomic power plant that would soon run on the fission power in uranium.

Now, Golovin said, "Sakharov has alerted us to solve a second, no less far-reaching problem of the twentieth century – how to produce inexhaustible energy by burning ocean water! That's a problem man could give his whole life to solve!"

Kurchatov scoffed, telling Golovin that he was getting carried away. Nevertheless, the men spent the next hour discussing how one could produce a successful hot plasma, and when it was over, Kurchatov had made up his mind. It was time the experimentalists took over from the theorists like Sakharov and Tamm and actually built such a machine.

"A huge operation will have to be initiated," Kurchatov reportedly remarked. "A problem for peace! Huge! Fascinating! We'll start the new year not with a weapon but with the Magnetic Thermonuclear Reactor and do a real job on it."

Bureaucratic approval of the plan, however, was not so quick, Golovin recalled many years later. The coordination of funding and industry orders dragged on for three and a half months and might have gone on longer had not one D.V. Efremov "extremely excited, run to Kurchatov with an issue of a magazine in his hand in which the results of Richter and the announcement of President Peron of Argentina were reported. Kurchatov immediately informed the administration of this and in the course of a week or two all our proposals were accepted and the funds were allocated."

In many respects, then, the Russian fusion work paralleled the effort by Lyman Spitzer – and efforts simultaneously under way at Britain's Harwell lab. Each nation was traveling a similar path, and the pace of research was virtually identical thanks to the sparking action of Peron's incredible speech.

Like Spitzer's Stellarator, the Sakharov-Tamm machine was a *toroid,* the Greek word for a hollowed ring such as that of a doughnut. The circular Soviet machine was girded with electromagnets at close intervals, like the Stellarator. But the Soviet machine included a second unique feature: a very strong electric current would be run through the plasma itself. Following the principles of electromagnetism, this current would cause the plasma to produce its own surrounding magnetic field. The plasma would thus be pinched by self-induced magnetism – at least in theory. The combination of the externally generated magnetic field and self-generated magnetic field would cage the plasma more evenly and completely.

Sakharov and Tamm called their theoretical device a toroidal magnetic chamber. The acronym in Russian was "tokamak."

The twenty-eight-year-old Sakharov turned over the design, then returned to hydrogen bomb work and never wrote about controlled fusion again. His sojourn into plasma physics had lasted less than a year, yet his ingenious idea for the tokamak propelled the Soviet Union into its own quest to recreate the energy of the sun.

The tokamak took on a life of its own, eventually populating laboratories all over the world. It would prove to be so successful that plasma physics, in whatever language, could hardly be discussed without the word *tokamak*.

Perhaps unsatisfied with the Tamm-Sakharov version of fusion's birth, many Soviet physicists offer another, more heroic tale. They credit the concept of magnetically controlled fusion to a nineteen-year-old soldier in the Red Army named Oleg Aleksandrovich Lavrentev. An inventive young man with no scientific training, so the story goes, Lavrentev sent a letter explaining his ideas about magnetic fields, electrical currents, and hot gas to the Communist Party's Central Committee in Moscow. The letter supposedly was passed on to Sakharov and Tamm, who worked up the formal scientific theory.

Lavrentev was later given a physics education and took a job at a leading laboratory.

It hardly mattered whether fusion in the Soviet Union originated with Tamm and Sakharov or Lavrentev. The notion was cropping up in many minds in many places, among scientists who attacked problems in the same manner. Just as Spitzer had recruited a team of young theorists from bomb projects, so too had Kurchatov. The laws of physics were universal laws, after all. But so were the laws of politics and secrecy. In the postwar era of blind competition, duplicated efforts were more common than the scientists knew.

In fact, the British were the first with a workable fusion idea, and they have proof. In 1946, Sir George Paget Thomson and Moses Blackman filed a secret patent application for a doughnut-shaped thermonuclear device they had designed at the Imperial College of Science in London.

Thomson, the son of the Nobel Prize winning physicist J.J. Thomson, had already won the physics prize himself. He had been involved in fission bomb research and had provided advice to the American effort at Los Alamos. He was also familiar with some of the American work on the hydrogen bomb. He was determined to push the British government to build an experimental fusion machine.

Thomson's investigations came out of a long tradition of historic discoveries by British scientists about the nature of the atom, the composition of stars, and the behavior of ionized gases – discoveries that led ultimately to the age of nuclear energy. It was logical that controlled fusion ideas should have been percolating first in Great Britain at the great universities.

Sir Ernest Rutherford, a British physicist who earned the Nobel Prize in 1908, might be considered the father of nuclear physics, laying the groundwork for that discipline. He evolved the nuclear theory of atomic structure and investigated radioactivity, showing how radioactive elements decay into other elements.

In 1919, at the Cavendish Laboratory in Cambridge, Rutherford performed an experiment crucial to the notion of fusion power and to the whole future of nuclear physics. He demonstrated for the first time the artificial conversion of elements, producing heavier elements through the collision of lighter elements in a process that would later come to be known as fusion.

Also at Cambridge, in the same year, the physicist Francis William Aston provided another critical building block of fusion science. He had invented the mass spectrometer, a device that can separate and measure atoms of different mass and for which he eventually won a Nobel Prize. With the spectrometer, Aston was able to demonstrate the existence of different isotopes of the same element. He also turned up the astonishing fact that the mass of a helium nucleus was significantly less than the sum of the hydrogen nuclei of which it was composed. This implied that the missing mass was somehow contained in the binding energy that held the nucleus together.

It took another famous British scientist of the time, the astronomer Sir Arthur Stanley Eddington, to make clear the implications of Aston's work. In 1920, in a speech to the British Association for the Advancement of Science,[4] Eddington raised the possibility that this energy locked in the atom could be used as an energy source. Eddington also speculated that the fusion of light elements beginning with the hydrogen-to-helium reaction was the process that powered the stars. He told his audience:

A star is drawing on some vast reservoir of energy by means unknown to us. This reservoir can scarcely be other than the subatomic energy which, it is known, exists abundantly in all matter; we sometimes dream that man will one day learn how to release it and use it for his service. The store is well-nigh inexhaustible, if only it could be tapped.

. . . Ernest Rutherford has recently been breaking down the atoms of oxygen and nitrogen, driving out an isotope of helium from them; and what is possible in the Cavendish laboratory may not be too difficult in the sun.

Some nine years after Eddington's speech, a German physicist, F. G. Houtermans, and the Englishman Robert Atkinson wrote a paper that is usually cited as the beginning of thermonuclear fusion energy research.[5] Trying to put Eddington's ideas correctly, they suggested exactly how thermonuclear fusion reactions could account for the energy of the sun.[6]

The theoretical work of a young Russian, Georgii Gamow, advanced the idea of fusion in stars even further. Under the laws of classical mechanics, a particle, in order to escape a nucleus, needed more energy than the binding energy restraining it. But Gamow, applying the new quantum physics and its probability theories to the problem, found something else: a theoretical means for a particle of relatively low energy to escape. This made it seem more likely that thermonuclear fusion might be occurring in nature.

Gamow discussed his ideas with John Cockcroft at the Cambridge lab. Presuming Gamow correct, laboratory machinery need not be so powerful as thought in order to create more nuclear transformations. The decision was made to build the Cockcroft-Walton particle accelerator, which soon confirmed Gamow's approach. In 1934, Mark Oliphant, Ernest Rutherford, and P. Harteck used an improved version of the Cockcroft-Walton machine to demonstrate the fusion of deuterium nuclei, releasing a significant amount of energy. But the machine was not useful for commercial power production, for it only fused one atom at a time.

It was against the background of World War II and its atomic bomb revelations that G.P. Thomson, steeped by his father in the new physics of the nucleus and of ionized gases, turned his thoughts to the construction of an actual fusion energy machine.

The thermonuclear fusion design Thomson patented used the same magnetic "pinch" principle applied by the Russians under Kurchatov's direction six years later. An electric current passing through the plasma would create its own magnetic field. Thomson's theoretical reactor would employ only this pinch effect to heat and hold the plasma. No other externally produced magnetic "bottle" would be necessary. The catch was that a tremendous current would be required, an expensive and difficult engineering proposition for which Thomson and his collaborator Blackman could not obtain government financing.

Still, for the next three years, Thomson, a physics professor at Imperial College, continually prodded Cockcroft, then head of Britain's new Atomic Energy Research Establishment at Harwell, to build the machine quickly on a power-producing scale. In January 1947, Cockcroft held a meeting at Harwell to discuss a controlled fusion program, and Thomson addressed the group which included the head of Harwell's theoretical physics department, Klaus Fuchs.[7] Thomson's far-fetched, ambitious proposal received considerable, hostile criticism and nothing resulted from the meeting.

As it happened, Nazi Germany had also thought of hot gases and self-induced magnetic fields. At the Harwell fusion meeting, the scientists discussed a report by an Allied commission that visited the German research labs following World War II. The report included brief notes about one M. Steenbeck. He had worked with a circular "pinch" machine, really a kind of particle accelerator he called the "Wirbelrohr," the whirl tube.

Thwarted in his own building plans, Thomson passed along the notes on the Wirbelrohr to two young doctoral students at Imperial College, Stan Cousins and Alan Ware, who decided to conduct some experiments.

In 1947, Ware devised a small experimental machine. It was a hand-built contraption made of old radar equipment and a glass tube. Ware was able to create powerful currents, and he observed a bright flash of light in the gas – just as Spitzer would in the Stellarator. But Ware could not come up with equipment that would measure the temperature of the plasma he was making.

The work was so preliminary and fresh that it was not yet considered secret in Britain. Ware discussed his work with James Tuck, a fellow British physicist who had just returned to Oxford after working with the Americans at Los Alamos on the atomic bomb.

Tuck later discussed the possibility of building a controlled fusion device with a young Australian, Peter Thonemann, at the Clarendon Laboratory in Oxford. Thonemann had come to Oxford the previous year in hopes of working on fusion devices as his doctorate. In 1939,[8] he had come up with a detailed concept for a fusion reactor when a student in Melbourne. Thonemann was directed to do his doctoral work on other nuclear physics, but he kept up the unconventional fusion work on the side, discussing it with Cockcroft, Fuchs, and countless others.[9] He and Tuck did eventually win some funding from the U.K. atomic energy authorities to investigate controlled fusion.

Tuck planned to build a "pinch" experiment, but before he could get started he received an invitation from Edward Teller to return to Los Alamos to work on the thermonuclear or hydrogen bomb, the H-bomb. Tuck jumped at the chance, but he did not forget controlled fusion altogether. A few years later, after the secret American fusion program had begun with Spitzer's Stellarator, Tuck proposed to the U.S. Atomic Energy Commission that he build a "pinch" machine at Los Alamos. In contrast with the gradiose name Spitzer had chosen for his machine, Tuck gave his research device a name that reflected more skepticism of the new science: the "Perhapsatron." In 1952, the Perhapsatron was added to the American program.

Thonemann, meanwhile, had begun a cautious investigation of elementary plasma behavior. He tried to answer some basic questions about the effect of electrostatic forces on the gaseous cloud before attempting the construction of an actual fusion device. His modestly priced efforts met with greater support from the British energy authorities than did Thomson's request to immediately build a powerful fusion experiment. Thomson's proposal was turned down repeatedly by Cockcroft, who was in the midst of founding an enormous research facility based on fission energy, not fusion.

As Thonemann and Ware were disseminating their first results, the curtain of classification descended on fusion research in Britain, prompted by a pair of spy scandals in 1950. Fuchs, a German-born British scientist who had worked with the Americans at Los Alamos before assuming a position of authority at Harwell, was convicted of giving fission bomb secrets to the Soviet Union. Another Harwell scientist, an Italian named Bruno Pontecorvo, was also discovered to have been a Soviet spy.

"I have little doubt," Thonemann wrote years later, "that Fuchs reported these matters [controlled fusion research] to his Russian masters who must have been aware of the U.K. interest."[10]

Thus, the timing of the Soviet Union's first foray into controlled fusion and the similarity of the tokamak to Britain's "pinch" machines might have owed something to the Fuchs connection.

The British government immediately took steps to tighten the security net around all atomic research, including controlled fusion. As a result, in 1951, the fusion research by Thomson and his students at Imperial College was classified secret. Ware's equipment was moved to a secure facility at the Associated Electrical Industries Laboratory at Aldermaston in Berkshire that had locked doors and bars on the windows. Thonemann's work at the Clarendon lab in Oxford was moved to nearby Harwell. There the Australian gathered steady support over the next decade and eventually built the devices Thomson had dreamed of.

Ware had managed to get two papers out before classification took effect. He would not be published again for the next seven years.

The scientists who turned to controlled fusion research in the United States, Great Britain, and the Soviet Union were the same men who, during World War II, had channeled their intellectual skills into the building of nuclear weapons. Because of this legacy, for some of them, the pursuit of peaceful fusion took on a certain redemptive meaning.

Kurchatov had been the organizer of seminal nuclear bomb work in his country. After Hiroshima and Nagasaki, he became increasingly concerned about the proliferation of nuclear weapons, as did his American counterpart, J. Robert Oppenheimer. By the mid-1950s, Kurchatov was campaigning for atomic disarmament and, at least, a ban on all nuclear bomb testing.

Kurchatov saw peaceful fusion research as a way to foster improved relations in the small nuclear club by bringing about a closer association among its scientists. He would call on the world's scientists to work together to "convert the energy of the synthesis of hydrogen nuclei from an instrument of destruction into a mighty vivifying source of energy, bringing well-being and happiness to all the people on earth."[11]

Like Spitzer, Kurchatov had come to appreciate that controlling plasma would take a long time. The research, he realized, was still in the stage of basic science after more than five years of secret work. The Russian began his effort to open the field by organizing a conference of Soviet scientists that reached outside his own Moscow laboratory. He then received Soviet government backing for his revealing 1956 speech in England. The initiative seemed to fit Khrushchev's political agenda, coming as it did just after the Soviet leader's famous address to the Twentieth Congress of the Communist Party in which he denounced Stalin and advocated an end to the Cold War.

At Harwell, Kurchatov spoke frankly to 300 of Britain's top scientists about a linear "pinch" experiment that had produced temperatures of 1 million degrees centigrade. It sounded quite positive at first. Kurchatov then revealed that his scientists had discovered something distressing about the neutrons emitted from the experimental chamber. They were not from true fusion collisions. Rather, the neutrons seemed to result from the separate acceleration of particles on the edge of the plasma.

Kurchatov did not mention Sakharov's circular tokamak, just his basic plasma physics theory. From a strictly scientific standpoint, Kurchatov had not divulged anything that could propel the West forward in fusion. It was all fairly tame.

"They weren't really giving away anything useful," said Ware. "They were just trying to get a propaganda ploy, I guess. It was part of the goodwill of that visit by Bulganin and Khrushchev."

Yet Kurchatov's speech held meaning far beyond its scientific content. As a political and diplomatic document, it was quite radical. In the United States, where the government's insistence on secrecy had prevailed over the scientists' protests, the speech was truly provocative.

By unveiling the existence of its own fusion research, the Soviet

Union was implying that there was no compelling military use for controlled nuclear fusion. It was also implying that the West was standing in the way of international cooperation in the pursuit of peaceful uses for the atom.

The speech, however, did not instantly inspire the United States to begin cooperating with the Soviet Union. One of the first effects was to stimulate greater interaction between the United States and its ally, Great Britain. Six months after the Harwell speech, a British delegation visited Princeton's Matterhorn Project for the first time, and in turn described work on the British "pinch" machine. Carefully worded papers on controlled fusion principles began to pass U.S. government censors and trickle out. Spitzer, tiptoeing around the fusion reactor question, produced a slim landmark volume on basic plasma physics, *The Physics of Fully Ionized Gases,* the first real textbook in the field. And he continued to press for declassification of all of fusion research.

Despite Kurchatov's revelation at Harwell and Spitzer's persistence, the chairman of the Atomic Energy Commission, Admiral Lewis Strauss, remained convinced that pursuing controlled fusion in secret was still in the best interest of the United States.

Strauss was a former Wall Street investment banker with a penchant for physics. He served during the war in the Navy's Bureau of Ordnance, where he learned about weapons production. He was also a special assistant to Navy Secretary James V. Forrestal, a fellow Wall Street banker. He liked to be called Admiral Strauss, even in civilian life. As head of the AEC and special adviser to President Eisenhower on atomic matters, Strauss was opinionated, ornery, and adamant in his warnings about the Soviet Union. He held the Cold War notion prevailing in the United States that the Russians were bent on world domination.

"Honesty, truth, and solemn covenants appear to be abstractions which have existence for them only as bait for those of other nations naive enough to believe in them," he wrote in his memoir, *Men and Decisions*.

Secrecy was vital to American security, Strauss believed, as was the continued stockpiling of weapons. He advocated the immediate production of the hydrogen bomb and supported controlled fusion power.

Spitzer recalled a visit by Strauss to the Princeton Matterhorn Project. "He talked about setting up a million-dollar prize for the first person or group that achieved the objective," Spitzer said. "He would say, 'Well now, if you had unlimited funds what would you do next? What could you do to accelerate the program?' It was an encouraging thing to hear."

Strauss complained that the Kurchatov speech at Harwell was just a ruse to get the United States to divulge valuable information. He was still concerned about the potential of a fusion reactor as a neutron source to make bomb fuel. At that time, he wrote in his memoir, "fissionable material – uranium – was still an excessively rare commodity and the key to the control of atomic energy."

But Strauss was fighting a rising tide. His own directors of research at the Atomic Energy Commission were advocating declassification. It was becoming clear that fission power plants could produce enough neutrons to breed bomb material and that fusion power plants were still a good distance away. Little by little, information was seeping out about fusion and there was a growing sense of the inevitable.

A few months after Kurchatov's speech, Spitzer happened to attend an international meeting of astrophysicists in Stockholm. The official topic was the role of magnetic fields in outer space. Among the Russian attendees was Lev Artsimovich, a stocky scientist whom Kurchatov had identified as his country's chief experimentalist in controlled nuclear fusion.

Spitzer and Artsimovich, both bound by their own governments' secrecy rules, skirted around the obvious subject. But when they sat down to an informal dinner, the Russian turned to Spitzer with a glass in hand. "He was making some comment about how he and I were each doing similar things under the wraps of secrecy," Spitzer recalled, "and he proposed a toast to plasmas."

"May they be worthy," said Artsimovich, "of the trust placed in them by the theorists."

The cryptic line held meaning for Spitzer. It implied that the Russians, too, were finding containment of plasmas difficult and unpredictable. The scientists could feel that declassification was imminent. As at other critical times in fusion's history, political events were needed to push the decision over the edge.

In the years to come, constellations of weather satellites, television satellites, and spy satellites would dot the night skies. In the 1950s, the skies were inviolate. Neither man nor machine had ever ventured beyond the earth's atmosphere. Rockets could not reach those heights, and a country separated by two oceans from its enemies could feel relatively invulnerable to missile attack.

That comfort ended on October 4, 1957, when the Soviet Union announced that it had used a rocket to launch a 184-pound satellite into orbit 560 miles above the earth. They called it Sputnik. Roughly translated, it meant a traveler. Radio signals broadcast by the satellite were picked up around the world as Sputnik ringed the earth, passing over the United States seven times a day. The American television networks broke into their regular programming to play the pinging sound of Sputnik's signal to a national audience. The satellite was even visible with binoculars at night.

Sputnik was an extraordinary feat of propaganda, in addition to being a technological first, and it snuffed all doubts about the Communist state's scientific prowess. The satellite's weight was eight times that of any on drawing boards in the United States. The launching also carried chilling confirmation that the Soviet Union had not made an idle boast the previous month when it claimed to have successfully tested an intercontinental ballistic missile.

At the end of the official announcement about Sputnik, the Soviet news agency Tass emphasized that people now could see how "the new Socialist society turns even the most daring of man's dreams into a reality."

To the West, the surprise Sputnik launching was part of a conscious effort by the Soviet Union to claim world leadership in science. The Soviet achievement shocked many American politicians. Democratic members of Congress attacked the Republican administration of Dwight Eisenhower for allowing the Soviet Union to be first in space and stage such an enormous propaganda coup. Recriminations were followed by calls for a redoubled American effort in rocketry. "For too many years," said Montana's Senator Mike Mansfield, the Democratic whip,[12] "the United States has been all too prone to underestimate the capabilities of the Soviet Union, and now the chickens are coming home to roost."

For Admiral Lewis Strauss, the Sputnik launching served to reaffirm his long-time warnings about the Soviets' scientific abilities. Skeptics, he said, had scoffed at his correct predictions that the Russians would develop an atomic bomb and, soon after, a hydrogen bomb. In a speech to a forum of nuclear physicists and engineers three weeks after Sputnik rose into the sky,[13] Strauss spoke about the need for a scientific alliance between the British and the Americans "in the interest of our mutual safety." And there was no area of the "scientific and technological cold war" more pressing, he said, than the development of peaceful atomic power.

Strauss saw the pursuit of scientific achievement as part of a global political contest between the Free World and the Communists, and he was passionate about winning that contest. He told the nuclear scientists:

> We must recognize the fact that, in some areas of the world, there are neutral peoples who harbor doubts as to whether we of the Western Democracies, with our political, intellectual, and economic liberty, can match the Soviet system of stark regimentation when it comes to making practical applications of scientific discovery. These peoples are waiting, some of them undoubtedly anxiously, to see whether the freedoms of the West can compete with the repressions of Communism in the development of the atom. Our system is on trial today as it has never been before.

The next field of battle would be the second United Nations Conference on the Peaceful Uses of Atomic Energy, set for 1958 in Geneva. Admiral Strauss needed a centerpiece exhibit that would advance the scientific reputation of the Free World and help dim the image of the orbiting Sputnik.

The first Atoms for Peace Conference, held in 1955 in Geneva, had been the initiative of President Eisenhower. In an attempt to lessen Cold War tensions, Eisenhower made several proposals in a speech to the United Nations urging international control of atomic weapons and the use of nuclear science for peaceful purposes. The major emphasis of the first conference was the potential uses of fission power for peaceful energy production. Fission powerplant research had been declassified by the United States, and Strauss's AEC was already authorizing commercial utilities to build fission reactors. The United

States, meanwhile, was not acknowledging that it had a fusion research program. That would change after Sputnik.

To prepare an impressive American presence for the Second Atoms for Peace Conference, Strauss turned to one of the nuclear projects under the purview of the AEC, the metal rabbit hutch that contained the promise of "the ultimate energy source for mankind."

Project Matterhorn and Lyman Spitzer's fusion machine would be enlisted in the scientific duel with the Soviet Union.

"As so often is the case," said Spitzer on Strauss's change of heart, "the desire for effective public relations is frequently an effective counter to the desire for secrecy."

It was time to open the door.

The scope of Strauss's public relations venture took Spitzer by surprise. The AEC wanted nothing less than an actual operating fusion device – a bona fide Stellarator – assembled in an exhibit hall in Geneva. And, to top it off, Strauss was demanding that the machine generate thermonuclear fusion neutrons, just the way a fusion power plant would.

Spitzer demurred. First of all, he argued, the Stellarator was not technologically advanced enough to generate what scientists called "thermonuclear" or fusion neutrons. It only produced stray neutrons from other interactions occurring in the machine, just as Kurchatov had indicated at Harwell. Although Juan Peron had claimed thermonuclear temperatures back in 1951, none of the world's plasma devices, so far as it was known, had made temperatures high enough and for a long enough period to create a stable, measureable, neutron-emitting fusion reaction.

Strauss's response to Spitzer was to the point: For Geneva, any kind of neutrons would do.

Secondly, said Spitzer, the cost and complication of moving the device would be enormous – perhaps several hundred thousand dollars. Even then, Spitzer thought, there would only be a fifty-fifty chance that the sensitive machinery would function properly at Geneva.

The AEC commissioners continued to exert what Spitzer felt was "considerable pressure" to exhibit a working, neutron-producing device. But eventually Spitzer's arguments were accepted. Instead, he proposed constructing a low-temperature device out of one of the working models at Princeton that would excite the hydrogen gas enough to show bright plasma encircling magnetic field lines. There would also be two "simulators" to illustrate Stellarator-style magnetic field lines. They would employ simple electron beams in glass tubes. In addition, the Americans would set up a large representation of the planned Model C Stellarator.

The exhibit's total cost would be $200,000 and would require 2,400 square feet of space. Strauss and the energy commissioners had been prepared to spend more for the sake of American scientific prestige. The proposal was approved.

The United States was ready to declassify controlled fusion research, and the AEC disseminated guidelines[14] permitting the scientists to publish "fundamental scientific work." The only caveat was that they could not divulge "the unique principles and technological details" needed to produce a sustained, continuous fusion reaction. In 1957, plasmas were commonplace, but sustained fusion was still a very distant dream. Thus, the effect of the declassification rule was to allow the Americans to publish and discuss virtually all their ideas. But they could not do so immediately.

In December of that year, the AEC sent a directive to the lab heads at Princeton, at the Livermore National Weapons Laboratory in California and at Los Alamos. Between then and the Geneva conference the following September, no matter what the temptation, the labs were to hold back on publishing news of any important advances in fusion, even if the Russians were publishing.

The British authorities had also agreed to hold back their fusion announcements – with one exception. In several months they intended to announce the results of a new experiment. Otherwise there would be silence for the maximum surprise effect at Geneva.

"We cannot have our cake and eat it, too," wrote the chief of the AEC's controlled fusion branch, Arthur E. Ruark.[15] "Our task is to overmatch the Russian papers and exhibits in September, if humanly possible."

On the eve of declassification, as preparations were advancing for the

Geneva atoms meeting, fusion research experienced one of the most sobering setbacks in its history, one that recalled the Peron incident and reinforced the axiom that rigorous science and public pressure do not mix.

By 1958, the state of fusion research methods everywhere was so rudimentary and the desire for success so high that the fatal combination of factors could have occurred just as easily at Princeton or Livermore or even Kurchatov's institute. As it was, it happened at the Harwell atomic research lab in England, where British fusion scientists were anticipating a real breakthrough in plasma control in a circular "pinch" machine they called Zero-Energy Thermonuclear Assembly – the Zeta machine.

On Saturday morning January 25, 1958, the newspapers of Great Britain exploded with bold, black headlines.

"THE MIGHTY ZETA," trumpeted the *Daily Mail* in inch-high letters. "LIMITLESS FUEL FOR MILLIONS OF YEARS."

And across the top of the page this ringing cry: "At last it can be told – Britain pulls off a triumph as great as the launching of the Russian Sputnik."

"BRITAIN'S H-MEN MAKE A SUN," wrote the *Daily Herald*.

"ZETA SPELLS H-POWER EVERLASTING," declared the *News Chronicle*. "Britain last night became – officially – the first country to prove that the H-bomb can be tamed to feed power-hungry nations. Harwell unveiled Zeta, its manmade sun, to show that we lead the world in Project H-Power Unlimited.

And in the *Daily Telegraph:* "U.S. ADMITS THAT BRITAIN HAS THE LEAD."

This fanfare of fusion triumph was generated by an unusual news conference called the day before by Sir John Cockcroft, the Harwell director. Before 400 reporters from around the world, the sixty-year-old Cockcroft, a Nobel Prize winner, took Britain's opening step into the era of fusion declassification. He announced that the hitherto secret Zeta had produced plasma temperatures of 5 million degrees centigrade and held those temperatures for up to three thousandths of a second.

It was his belief – he was 90 percent sure, he said under hard questioning – that thermonuclear reactions were taking place. Zeta had apparently created a controlled fusion reaction, the world's first.

Later, on British television, Sir John was ebullient. "To Britain this discovery is greater than the Russian Sputnik," he declared. Cockcroft had plans to boost Zeta's 5-million degree plasma to 25 million, and he said he believed British industry would have a working fusion reactor in twenty years' time.

Physicist Alan Ware, who helped analyze Zeta's findings, recalled in an interview nearly three decades later that the political bureaucracy played a key role in the affair.

"We were certainly very encouraged by the results," Ware said. "They were much better than we'd had previously. The discharges were much less unstable. The fluctuations were much smaller and we were getting the right number of nuclear reactions for the temperatures we thought we had. We reported to Cockcroft, and he was of course excited. And then they passed it up the chain to the politicians. Britain badly needed some good news at that time, and I think they were the ones that decided they'd publish it. There was a general greater enthusiasm for publishing the further you went up the chain."

The British news media blazed with chauvinism. Newspapers printed huge pictures of the 120-ton machine – the largest fusion device in the world at the time – and of the scientists who built it. The builders included Peter Thonemann, the young Australian who had won Cockcroft's support for a small machine a decade before and then led him by enthusiasm to big-scale science. "NOW MEET THE AMAZING YOUNG MEN WHO ARE BRINGING IN THE NEW AGE OF POWER," wrote the *Daily Mail* above pictures of the smiling researchers.

The news also quickly reached the Soviet Union. Moscow Radio reported the Zeta claims while pointing out, "as we all know," that the Soviet academicians I.E. Tamm and A.D. Sakharov in 1950 had first proposed the use of magnetic fields for controlled thermonuclear devices.

At Kurchatov's institute, Lev Artsimovich and his research team rushed to have the official scientific paper on Zeta translated from that week's issue of the journal *Nature*. At first, they were incredulous. Artsimovich was a careful, skeptical scientist whose first reaction to most news of success was "Chush Cobachi!" – dogshit. But the British report convinced the Soviet hierarchy that its fusion researchers needed their own experience on this toroidal "pinch" machine. They could not afford to be left behind if the British machine really had worked. Within a few months, a Russian copy of Zeta was under construction – but not under the direction of the still skeptical Artsimovich.

As soon as the British announcement flashed to newsrooms in the United States, Strauss vigorously denied to reporters that the Americans were behind in the fusion race. In fact, the U.S. government had arranged with the British to disclose results from two Los Alamos "pinch" machines the same day as the Zeta findings. The *New York Times* paired the announcements in one article and although the Los Alamos results were slightly weaker than Zeta's, the *Times* concluded that the two nations were "neck and neck" in the race for fusion.[16]

The *Nature* issue, in addition to publishing the Zeta and Los Alamos results, also contained a short, polite article by Lyman Spitzer, rebutting the British findings. There was a real contradiction, Spitzer wrote, between what theory said should be happening inside Zeta and the rate of particle heating the Harwell scientists were reporting. The heating was just too fast to be explained by fusion collisions.

"Some unknown mechanism would appear to be involved," Spitzer wrote. He could not say what exactly was happening inside Zeta, but to his mind it was not the fusing of hydrogen nuclei.

At California's Livermore lab, two highly talented and irreverent young physicists, a graduate student named Harold Furth and a recent PhD named Stirling Colgate offered more clamorous criticism of Zeta than did Spitzer. "Lyman was far too polite to say this was all bunk," recalled Furth, who later became director of the Princeton laboratory. "Colgate and I were in the rebellious period. We weren't polite at all, so we made a terrible stink."

Britain's fusion triumph proved to be brief. After a few more months of experiments and calculations Cockcroft issued a sober news release in May. The neutrons produced by Zeta were not coming from true fusion reactions. They were the result of a reaction that was a byproduct of plasma heating. It was clear that these highly excited runaway particles did not reflect the overall temperature of the plasma. The plasma, it turned out, had not reached the high temperature necessary for fusion.

It would be another decade before the world's fusion scientists

would devise equipment that could reliably tell them what was happening to the plasma they were creating.

In articles here and there, the international press took note of Cockcroft's corrected results. In Paris, *Le Monde* had this June headline: "CONTRARY TO WHAT WAS ANNOUNCED SIX MONTHS AGO AT HARWELL – BRITISH EXPERTS CONFIRM THAT THERMONUCLEAR ENERGY HAS NOT BEEN 'DOMESTICATED.'"

The world's plasma physicists took no delight in the British humiliation. They all shared high hopes for fusion's success and now those hopes had been dashed once again – in full view of an international audience. Declassification had begun with an excited and ultimately imprudent display. The fusion scientists looked foolish now. Working without the cross-checks and healthy skepticism of peers from other laboratories, the British scientists had fallen victim to their own optimism and to the political pressures of the times.

Zeta, "greater than the Russian Sputnik," proved to be a useful research machine for the next ten years. But it would always be remembered as a British debacle, one of the broken promises that would litter the path of fusion development.

3
Friends and rivals

The first meeting of the world's fusion researchers was an exercise in cautiously scripted emancipation on the shores of Lake Geneva. Sponsored by the United Nations, the 1958 Atoms for Peace conference was the largest international gathering ever to focus on the potential for taming nuclear energy for peaceful purposes. The promised revelation of secret fusion research by the United States, Great Britain, and the Soviet Union was the main attraction, and 5,000 scientists, government officials, and other observers came to bear witness.

The conference unfolded over two weeks in September in two venues. One was the assembly room of the old League of Nations building, where delegates from sixty-seven countries heard formal speeches from the world's top nuclear scientists. The social arena was a vast exhibition hall constructed for the conference as a kind of nuclear world's fair. It was open to the public and packed with displays about fission and fusion power. To aid communication, female linguists from a Swiss translation school were recruited as guides. Dressed in pert, dove gray suits, with matching berets and short white gloves, the young women took up positions around the hall as thousands of curious Swiss and other international visitors streamed through.

The conference became a duel for world leadership in fusion even before the official opening speech of United Nations Secretary General Dag Hammarskjold. On the weekend before, Admiral Lewis Strauss, who was chairman of the U.S. delegation, and Britain's Sir John Cockroft, held a joint news conference in Geneva announcing that their two nations were ending all secrecy in controlled fusion research. On the same day the Soviet Union revealed that it had built the world's largest fusion research device to date at the Moscow institute run by Kurchatov. On the opening day of the conference the Russians made their declaration of declassified research as well.

The Americans had come to Geneva with high expectations, great curiosity, and not a little fear about what progress the Russians might have made in fusion. There were rumors that the Russians had solved the riddle of plasma heating and actually had plans for a prototype fusion reactor. Some people worried that the Russians would show up with miniature reactors.

"There was that feeling that we were losing the race," recalled Ed Meservey, one of Lyman Spitzer's physicists from the renovated rabbit hutch. "It was the time when Sputnik went up. We were losing the space race; we were losing the fusion race."

The large Princeton contingent included Spitzer, his deputy, Mel Gottlieb, and Don Grove, a young physicist with coal black hair and an aquiline profile whom Spitzer had borrowed from Westinghouse Electric Corporation to design a hypothetical reactor right down to the engineering specifics. Grove stayed on at Princeton to oversee the planning and construction of the Model C Stellarator. An eager, competitive person, Grove relished the chance to match achievements with the Russians.

California's Livermore National Laboratory sent Edward Teller, who was to give a major speech on the U.S. fusion program, which had grown tremendously in the years since the first Stellarator. With 432 Americans as delegates, the size of the U.S. contingent to Geneva was proof of that.

The American exhibit fulfilled Strauss's demand for impact. It took up close to half the hall. Meservey and a crew of American workers had spent the whole summer in Geneva setting up the working model of the B-Stellarator and the other machines. Los Alamos scientists brought over a Perhapsatron "pinch" machine that made millions of nonthermonuclear neutrons every second. A few days before the exhibit the Russian scientists were allowed a preview. Grove and Dick Post from Livermore both remembered how the Russians' jaws dropped at the sight of the enormous American exhibit, with its glowing plasmas in

Britain's ill-fated ZETA (Zero-Energy Thermonuclear Assembly), 1958. *Courtesy UKAEA.*

"To Britain, this discovery is greater than the Russian Sputnik."—SIR JOHN COCKCROFT.

working devices that were nevertheless not quite genuine fusion machines.

"They were really flabbergasted," said Post. "Then we watched the wheels go around." The Russian response was to beef up their exhibit. A day or two later trucks started arriving with additional Soviet displays. The Russians too had a working machine – a straight line "pinch" device – that made plasmas every so often with a tremendous burst of sound. The Russian exhibit also featured a highly polished full-scale model of Sputnik III – "which of course had no bearing on the theme of the conference," Admiral Strauss wrote later. "It was a bare bid for prestige and interest, without apology, and it succeeded – at least as far as the general public was concerned."[1]

Boast followed upon boast. In the fission energy field, the Russians provided some drama by showing a movie of a newly operating 100,000-kilowatt fission power plant in Siberia, then the largest nuclear power plant in the world. Until they viewed the film, Western scientists had no inkling of the plant's existence. The largest U.S. power reactor at that time was the Shippingport, Pennsylvania, reactor, rated at about 60,000 kilowatts.

The spirit of rivalry and bravura was keen and palpable in the exhibition hall. Geneva was to be a celebration of openness, a gathering in which the scientists could speak and meet with an ease denied them for so many years. But Cold War reflexes of suspicion and competition were not to be dispelled by an assurance of declassification.

Grove recalled years later how such paranoia was fed. An evening boat trip on Lake Geneva sponsored by Westinghouse was a social highlight. Some of the attractive female translators were invited to mingle with the virtually all-male spectrum of scientists, and a handsome, dark-haired Russian guide attached herself to Grove for the cruise. His enthusiasm dimmed, however, when word was passed to the Princeton scientists that some of the guides might be spies and to take precautions.

Coached to be prudent about information, Grove became even more suspicious later in the conference when Wolfgang Stodiek, a West German scientist, repeatedly visited the full-scale model of Princeton's as yet unbuilt C-Stellarator. Stodiek pored over a four-inch-thick proposal book that Grove had brought along as a reference. After several such visits by Stodiek, Grove began to worry about how much information should be revealed on an American machine that was still on the drawing boards. He discussed his anxiety with Mel Gottlieb, Spitzer's deputy, and they decided to make the proposal book "unavailable." When Stodiek showed up for another look, Grove walked behind the display where he usually kept the book and then reported that it had "disappeared." The persistent Stodiek returned several more times, but the lost book never surfaced.

"Maybe the Russians stole it," Grove shrugged.

Stodiek remembers that he felt Grove was really implying that the German had taken it, but Stodiek kept his anger to himself.

Just a year after the incident, as new fusion alliances blossomed,

Stodiek came to work for the Princeton Laboratory on that same C-Stellarator and eventually collaborated with Grove on more than fifty papers. One late night as they were operating a C-Stellarator experiment and reminiscing, Stodiek turned to his good friend Grove and asked, "Did you ever find that book? I was sure the Russians took it." With some embarrassment Grove confessed, explaining what the paranoia of Geneva had led him to do before international trust had come to the fusion program.

The mistrust of 1958, however, was not entirely baseless. At the Geneva conference, Dick Post was among those who experienced a lame bit of espionage firsthand. Years later, he told the story this way: He was eager to speak freely at last about his experiments on a straight-line magnetic device that was a promising alternative to the toroidal Stellarator. To prepare for his speech, Post had annotated one copy of his paper, marking it in the margin with red ink to denote five minutes, ten minutes, fifteen minutes, and so forth to pace himself to finish in the allotted time. He placed the red-marked copy in his briefcase and went over to a rehearsal session at the United Nations headquarters.

At this organizational session, the international group of scientists congregated on a terrace overlooking the city and talked animatedly about their papers. After awhile they got up from their chairs to go inside for some soft drinks. Post left his briefcase on the terrace. He returned after about five minutes, picked up his briefcase and then went back to his hotel. When he looked for the annotated copy a little later he could not find it. Puzzled, he searched around the room for a bit before shrugging off the loss. He annotated another copy and gave the speech as scheduled. The audience was given reprints without the margin notes.

The next day a little Russian scientist, whom Post characterized as "innocent as the driven snow," came up to him in the conference hall. Clutched in his fist was a copy of Post's speech – with red-inked numbers in the margin. The Russian wanted some help. "I don't understand this and this," he said, pointing to the text.

Post was shocked to see the missing annotated paper and then, in turn, greatly amused. It was obvious that the diminutive Russian had no idea the paper he was holding was Post's missing original. The annotated paper had made its way to the Soviet delegation, presumably after a security agent had dipped his hand into Post's briefcase on the terrace.

The paper, a soon-to-be public document of no extraordinary significance, had ended up in the possession of this unsuspecting Russian scientist.

It was hilarious, Post remembered thinking, but he kept his deductions to himself, answering the Russian's physics questions with a straight face in the spirit of scientific detente.

On the second day of the conference the fusion scientists crowded into the League of Nations assembly hall. This was the moment of reckoning. The first day had been spent mostly on formalities. Now, in major speeches, each country's top scientists were set to present the first broad revelations about what they had achieved in fusion.

Lev Artsimovich, the Soviet Union's tough and skeptical machine builder, had written the keynote address. A colleague delivered it. The Americans and the British listened with trepidation to the translated text as the speech moved through intricate plasma theory to a description of the actual Russian experiments. In the exhibition hall everyone had been impressed by the dimensions of the new Russian fusion machine called OGRA. Photos on display showed scientists dwarfed by the sixty-four foot long linear pipe. How close to a reactor was the giant OGRA? How far behind was the West?

As the audience of national fusion fraternities absorbed the Artsimovich speech, it became apparent that OGRA was just an experimental device, not a reactor, and the Russians had no more answers than anyone else. The Soviet Union had built a wide array of devices, but, wrote Artsimovich, "not a single one of these approaches has been explored to such an extent as to permit one to say that success is assured."

The Russians seemed completely open at the conference. They presented the conference with a four-volume set of papers summarizing the Soviet work of the past seven years. The books included the seminal Sakharov and Tamm papers on the device called the tokamak.

What most stunned the Atoms for Peace participants was the uncannily similar course that fusion research had taken in each country, despite the almost complete isolation that had existed until 1956. As 104 unclassified speeches on controlled fusion tumbled forth, it became clear that the United States, Great Britain, and the Soviet Union all had invented circular and linear devices using magnetic principles. None had a working fusion reactor. All had experienced uncontrollable plasma instabilities in every machine. All had arrived at similar points in their theoretical calculations of what physical requirements were needed to create a power-producing reactor.

In short, all three countries had been stumped by the plasma. None had the answer to controlling nuclear fusion. To the scientists, it was both a relief and a disappointment.

"We realized the bloom was off the rose," said Post, "because one of the consistent themes throughout the conference was that the kind of thing that happened to Zeta – instabilities – were really turning out to be a major problem." The idea that instabilities could easily be solved, he said, was "too naive."

Quoted in the *New York Times,* Dr. H.J. Bhabha, chairman of the Atomic Energy Commission of India, said that the similarity of approaches and findings arrived at independently was "an indication of what secrecy has cost the world for no useful purpose."[3]

In their speeches both Artsimovich and Teller addressed the issue of secrecy and the uncertain state of fusion research. Both men sounded somber notes about the possibilities for quick fusion success.

"Of course I fully realize that at the present stage of our knowledge any discussion of (future thermonuclear reactors) can only rest on our faith in the ultimate triumph of human ingenuity," Artsimovich wrote. Considering that the world's scientists had no better than a basic understanding of fusion, the Russian concluded that the true value of the Atoms for Peace conference was the simple fact that his colleagues were meeting.

"For the first time these results will be discussed on an international scale, and this is probably the most important step which has been made towards the solution of this problem," Artsimovich wrote. "The importance of this fact is greater than that of the separate investigations, which as yet have not brought us very much nearer to our ultimate goal."

Teller expressed similar sentiments but his pessimism about the near term success of fusion energy research was more pronounced, because he had an eye on economic factors. The problem, he explained to his audience, was that the intricate machines needed for fusion power production, coupled with the internal radioactivity of the reactor, would "make the released energy so costly that an economic exploitation of controlled thermonuclear reactions may not turn out to be possible before the end of the twentieth century."[4] It was a note that would be sounded with more force in coming years.

Although the Atoms for Peace conference lasted just two weeks, in the exhibition hall and in the restaurants and cabarets of Geneva, a profound change was taking place among the fusion scientists. They had been schooled to compete, but they had a tremendous amount in common – an idealistic goal, a daunting intellectual problem and, as they learned at Geneva, a parallel history of frustration. In a rush of recognition and relief, the national groups of researchers coalesced into an international body. The Geneva conference marked the birth of the world fusion community, and from that time forward fusion curiously became almost a sacrosanct kind of cooperative endeavor for both the scientists and their government backers.

Over the ensuing four years political dramas would shake East – West relations. Russia would build the Berlin Wall. Khruschev would be faced down in the Cuban missile crisis. But cooperation in the obscure field of fusion research would proceed quietly and cordially, even if a residue of competition lingered.

The unwritten rules of the game had been drastically loosened, but there were still rules. The nations would share willingly, so long as no one had a clear advantage. There would be no secrets, at least until the basics of fusion physics were solved. When commercialization came, however, the doors would be shut again and economic competition would begin. That was understood. In the meantime, the scientists could talk, write, and visit one another. The men in the Princeton rabbit hutch were free at last, and within a few months of the Geneva conference two of them had broken the precedents of the past and made their way to Moscow and Kurchatov's door.

Mel Gottlieb, Spitzer's tall and gregarious deputy, and Ed Frieman,

Spitzer's top theorist, were eager to go to Moscow. At Geneva they asked the Russians and, in the friendly spirit of the moment, were promised an invitation.

It took several months and a series of cables to extract the formal invitation, but in the early winter Gottlieb and Frieman became the first Americans to tour the Kurchatov Institute. (England's Sir John Cockroft had made the trip about a month before.) When they arrived, said Gottlieb, workers were "hanging out the windows" to get a glimpse of the foreign visitors.

"It was a huge facility," Frieman recalled years later, "with much more going on than we were allowed to see. We were carefully shepherded around from one place to another through certain doors. What impressed me at the time was how primitive their equipment was." He gave an example. "You would see an oscilloscope, clearly quite old, heavily constructed as if built for a battleship. It was impressive they were able to make the strides they had, in view of the equipment they were working with."

In addition to the guided tour, the two Americans were able to have private discussions with the Russian physicists, many of whom spoke English. "They were just as hungry for information as we," said Frieman. The subjects ranged from the scientific and esoteric to the intensely personal. "They talked of their families being wiped out by World War II, showing us scars from the war, very human things, and they had an extreme curiosity about life in the United States. They asked me, 'What kind of house do you live in?' and I said it had two floors. 'Which one do you live in, top or bottom?' I told them 'both.' 'How could that be?' they asked."

The trip to the Kurchatov Institute was a rewarding personal experience, Frieman said, but it did not yield a great deal of new scientific information. It only confirmed what they had learned at Geneva. "The people we met were very bright," he said, "but when looking at basic concepts we really found we'd done much the same sorts of things. We were pretty much following a parallel course."

The British were also cultivating interaction with the Russians. In November 1958 Cockcroft was the guest of the ailing Kurchatov, who had suffered several strokes and intermittent dizzy spells that kept him from the Geneva conference. In 1959, Cockcroft invited a Russian group to tour the British facilities. In May 1960, a Russian group toured the U.S. fusion laboratories. An American group made a return visit in July.

The contacts went beyond polite snooping. In 1961, Great Britain and the Soviet Union signed a five-year pact to cooperate in nuclear energy research. Shortly thereafter, the United States and Great Britain formally extended the cooperative pact they had made during the era of secrecy.

Meanwhile, the lifting of classification had immediate effects on each nation's domestic program. The British, who had been pursuing fusion at the Harwell research lab, decided to build a separate, unclassified plasma physics laboratory at Culham, outside of Oxford, providing the scientists and their visitors with a less inhibiting environment.

In the United States, recruiting young scientists to fusion was suddenly easier. Princeton University, with a Ford Foundation grant, immediately began a plasma physics graduate program. "We could finally tell our colleagues in the physics department what we were doing," said Frieman. "Plasma physics became a recognized intellectual discipline in the mainstream of physics. There was suddenly a new infusion of ideas, a new infusion of people."

Dick Post wasted no time in spreading the fusion message. Barely a month after Geneva, he left his desk at the Livermore lab and showed up in New York at the annual meeting of the American Physical Society. Hundreds of physicists were at the convention. No longer forbidden to speak of his program, Post intended to educate, to inspire, and to convert some fresh, sharp minds to the vision of a fusion future.

One listener, in particular, was profoundly affected by what Post had to say, and his memory of that time is vivid. The story of Robert Gross's conversion is but one illustration of how powerful the fusion message was for scores of young physicists.

When he attended Post's lecture, Gross was twenty-six years old. He considered himself "pretty hot stuff" in the rocket industry, which was booming thanks to Sputnik. Gross had his own laboratory at Fairchild Engine and Airplane Corporation in New York. "For a young kid I had a lot of responsibility," he recalled. "I didn't realize I was unusual. I thought that was the way it was. Generals used to come – I'd need another half a million dollars and, bango, it was there. Those were glorious days. The government was very magnanimous then to anyone who had good ideas, good credentials."

Gross had a great setup in the rocketry world, but he was about to be touched by Post and the mission of fusion.

At first, Gross could not understand anything Post was saying. The jargon of fusion physics that had evolved under classification was incomprehensible. Still, Gross could make out that fusion was "something way beyond anything I had ever thought about." He had viewed himself as a high-temperature physicist, dealing with engine fires of one or two electron volts. But Post was talking about hundreds of thousands of volts for controlled fusion machines.

After the meeting, although Gross searched eagerly for books about nuclear fusion reactors, he found none. Classification had done its job well. Gross was relieved and excited when an invitation came from Edward Teller asking Gross to visit the Lawrence Livermore National Laboratory in California with about fifty other potential recruits to learn more about plasma physics and peaceful fusion power. In an old movie house in Livermore, the young man sat for three days listening to Post, Teller, and others describe the promise of fusion. Gross still did not understand much because of the jargon, but he was mesmerized.

"I came home all excited, paced the floor, thought maybe it's all just a fad." But Gross was hooked. He sold his house, took a leave from his rocketry job, packed his family in an old Ford stationwagon, and made the journey west. After a year studying fusion at Livermore, Gross made his choice. Years later, he said:

> "For reasons that I don't fully understand even to this day, I decided that I was going to devote my life to it. The idea of bringing a whole new energy source to mankind, civilization, I said that's something worth devoting my professional life to. It was so very idealistic. Everybody in the business is very idealistic. The fusion bomb was a terrible thing, but gee, it's also a great thing if we can control it. No more energy shortages . . . it would be the solution for all mankind forever, and I really believed that then, and I believe that now."

Within a year Gross became an East Coast missionary and started a fusion research program at Columbia University in New York. His program would produce crop after crop of sterling graduate students for the cause. He would become dean of Columbia's School of Engineering and Applied Science and he would sit on White House committees on fusion and science policy. Like Post, his initial idealism undiminished, he would spend decades leading the crusade.

In such a manner did the fusion ranks swell after the lifting of classification. Fusion was a powerful and seductive idea, and scientists who could envision a fusion future took to it like a religion. Geneva had given them the freedom to deliver their message. With a new strength in numbers, perhaps they could overcome the hurdles to a fusion reactor.

4
Searching for answers

The fusion researchers had begun their quest with enormous confidence, but the decade after Geneva evolved into a seemingly endless period of uncertainty and groping. There was trouble on all fronts.

The test reactors coughed out confusing results, and the men who ran them were at a loss to explain why. On an engineering level, plasma temperature remained difficult to measure accurately. In terms of concrete progress, none of the machines – now a worldwide menu of devices – could contain plasma for more than several thousandths of a second, if that. Even more disturbing, the idealized theory upon which the entire field of plasma physics was based appeared increasingly irrelevant to what was really happening inside the machines.

In some respects, Geneva had merely transformed private hope into public frustration, played out before an international body of peers. Instead of just debating among themselves, the national teams of scientists were compelled to defend their ideas at the periodic "Plasma Olympics." The first international gathering of the fusion researchers following Geneva was marked by a particularly acrid debate over reactor measurements in which Western scientists received their first lesson in Russian-style scientific criticism from the formidable Artsimovich, leader of the Russian experimental team.

Lev Andreevich Artsimovich was one of the great men of fusion and, by all accounts, one of the great arguers of his day. He died in 1973, but stories about him abound, and his caustic aphorisms are frequently quoted with reverence and amusement by international colleagues.

Photographs invariably show a stocky man standing with his feet

Sebastian Pease, one of Britain's earliest fusion researchers and later director of Culham Laboratory. Courtesy UKAEA Culham Laboratory.

"It was a real physical challenge which it seemed to me we shouldn't walk away from. Quite a lot of my colleagues thought we were mad."—SEBASTIAN PEASE.

wide apart, a boxer's balanced stance, his thumbs in his pockets. He had a broad, smooth face and, although he smiled for the cameras, people who knew him use the darker adjectives to describe his personality. He was grim, brusque, brash, outspoken, and furiously competitive. He was "a man who would not brook opposition," observed Mel Gottlieb, the Princeton physicist. "Artsimovich was very hard on his own people, rarely a kind word."

Artsimovich is also described as intellectually honest, straightforward, and possessed of a sharp wit. He is remembered best for the challenges he put to his newfound brethren who were unaccustomed to his pugnacious approach to scientific exchange. The fact that he had been a boxer in his youth explained a good deal.

His most famous intellectual sparring match took place during his international debut. Artsimovich had been absent at Geneva, although his speech had been read there. Three years later, in the fall of 1961, the United Nations' fledgling International Atomic Energy Agency (IAEA) organized its first conference on controlled nuclear fusion in Salzburg, Austria. It was the inaugural "Plasma Olympics" and Artsimovich was in top form.

The topic that touched off the sharpest debate was a new sort of fusion device, not the doughnut form but an alternative reactor called the magnetic mirror machine. Both the Soviet Union and the United States had been spending considerable money and time on the device. The Russians' giant OGRA pictured at Geneva was a mirror machine.

The concept was rather simple. It consisted of a straight section of tube belted with coils that produced varying magnetic fields inside. To prevent the speeding plasma particles from shooting out the ends of the tube, the coils were arranged and operated so that the magnetic field in the center was weaker than those at the ends. The strong fields at the ends were supposed to repel or reflect the plasma back in on itself. Thus the name "mirror" machine.

Theory showed the concept to be plausible, but the actual machines did not work well. In fusion parlance, they were leaky. Too many of the fast plasma particles, excited by tremendous heat, escaped out the ends of the invisible magnetic bottle. That was hardly alarming. Other fusion devices leaked, too, only out the sides.

In the United States, the mirror machines had formed the main controlled fusion program at Livermore. Fred Coensgen, often referred to as "poor Fred Coensgen" when this story is told, presented a paper in Salzburg on the latest results from Livermore. Lev Artsimovich was in the audience, ready to pounce.

A preliminary version of the paper had been published before the session, and when Artsimovich went over it he found that Livermore's results differed from those of his own mirror team. The Americans were claiming that they had been able to hold onto a plasma in the small mirror machine called Toy Top for one millisecond, one thousandth of a second, a real achievement for the mirror concept. In studying the Americans' methods, the Russian discovered an apparent mistake in the way the Americans were counting the neutrons thrown off by fusing particles in the plasma. A device on the machine counted the neutrons, but the Livermore scientists had failed to take into account the fact that the epoxy glass protecting the measuring device could also slow down the neutrons. Neutrons thought to be emerging from the reaction one millisecond after it began probably had emerged much earlier and were just being counted late. This implied a longer fusion reaction than really existed.

Before they arrived in Salzburg, Coensgen and his team had realized the error. And in the oral presentation of the paper, he corrected it. That did not prevent Artsimovich from making a public show of the Americans' blunder.

When the floor was opened to questions, Artsimovich was unmerciful. According to the IAEA's translated record of the debate, the Russian unleashed a hail of sarcastic blows to Livermore's neutron figures, saying that the longer neutron emission was due to the effect of the epoxy measuring device. "This is not very surprising," he added, "since anyone who has been using such counters knows well that there always is a background of slow neutrons."

"Coensgen was absolutely slaughtered by Artsimovich on the measurement error," recalled one witness, the Briton Roy Bickerton. "It was a memorable experience, this stocky Russian figure who we didn't know too well standing in the aisle and really giving it to him."

But Artsimovich was not through. Later in the session he rose to attack a general paper on the Livermore mirror program given by Dick Post, U.S. inventor of the concept. The results arrived at by Artsimovich's researchers, said the Russian,

> are in sharp contradiction with the attractive picture of a thermonuclear Eldorado which has just been drawn by Dr. Post. After the initial assertions of Dr. Coensgen that the plasma containment time was about one millisecond have proved erroneous, we now do not have a single experimental fact indicating long and stable confinement of plasma with hot ions within a single magnetic mirror geometry.

Artsimovich's diatribe might have been called boorish or brutally honest. The American mirror program, so far, was a bust, he was saying. Face it.

The Americans bristled, and a few rose in the audience to defend the American program during the short time allotted for commentary to the Livermore paper. Mel Gottlieb, always the statesman, proposed a special evening session to work out the disagreements. Later in the conference, Artsimovich's researchers gave evidence of a way to make a more stable plasma in the mirror machines, evidence that seemed the more impressive for the Americans' embarrassment. It felt to some like a setup.

"We discovered, as we got to know the Russians, that it was more the style there to make very strong attacks in public," said Marshall Rosenbluth, the chubby theorist whom Artsimovich later dubbed "The Pope of Plasma Physics." "What looked to us like shockingly rude behavior was normal for them."

Years later, Coensgen recalled the incident with some lingering bitterness. "This was a little bombshell he wanted to drop and make us look bad, make us look like idiots," he said. "I hadn't really expected international politics would enter into this thing."

After the pasting in Salzburg, Post and Coensgen returned to Livermore still dedicated to making a success of the mirror machines. They adopted the Russians' modifications. For Post, a man of unrelenting optimism, it still was a time of high hopes.

"We thought it would take just one clever idea," said Post. "We were naive."

Mirror machines survived into the 1980s and Post was a survivor as well, his enthusiasm for fusion as undiminished as it was for mirror machines. After retiring at sixty-five, Post still kept an office at Livermore and showed no signs of fading from the field. Around his eyes he had the crinkled crow's feet of a man whose most natural expression is a smile. He retained a belief that all scientific problems, indeed any problem, would eventually give way to reason. "I'm not a handwringer," he said. "I have in spades the male fallacy that you can do something about everything. If there is a problem, I have to think of an action."

The most telling story he offered about his own character had to do with the assassination of John F. Kennedy. To cope with his grief, horror, and shock over the news of the president's death, Post said he tried to think his way out of the problem. He ended up inventing a method of protecting an open car with a radar device that would instantly put up a screen against oncoming projectiles. He sent the plan to the FBI. It was promptly classified secret.

The mirror machine was Post's child. He had been working on the magnetic mirror problem at Livermore since 1952, the year after Lyman Spitzer started the Stellarator program at Princeton. The new Livermore national lab was just coming into its own then, created with the help of Edward Teller to compete with the Los Alamos national lab. Besides bomb work, Teller wanted to include some controlled fusion research.

Herbert York, the lab director, proposed offering an alternative fusion scheme – a device that did not close in on itself as did the Stellarator. York reasoned that a straight-line machine, if it could be devised, would be much more simple in its engineering, might work better without the complicating curves, and would therefore be more economically attractive as a reactor.

Post was assigned to investigate the magnetic theories that might apply to a straight-line device and to figure out how to plug up the ends.

"The magnetic mirror effect was known from cosmic ray theory," Post explained. "A Mexican physicist Vallarta had done some very nice work. Cosmic rays come zooming in from outer space, and they come in contact with the earth's magnetic field and then they get turned around by the earth's field and bounce back off. That was a starting point conceptually."

Post was also eager to offer an alternative to the toroidal devices. "Fusion is too big an objective to think you're omniscient about how it's going to be solved," he said.

In the era of international secrecy there was a feeling of imminent discovery. Post's wife Merilee remembered, with some nostalgia, the tension and naivete of the early mirror machine years. Back then, she said, whenever she left her house to work in the garden she would pull up the window and leave the telephone on the ledge so she would not risk missing the call "that would be Dick saying they'd got neutrons or not. We were waiting for neutrons. It was exciting to think that very afternoon there could be a breakthrough to prove fusion."

At dinner, when Post would come home from the lab and describe his work to his wife, she would pay special attention. She worried, she said, "If he got struck down, would I be able to remember enough of it to tell someone – the one idea – the secret formula?"

Just one clever idea was needed, Post had said. After Geneva there were many people working on the problem, but no secret formula emerged. In the first decade of open research, cleverness seemed to reside more with the plasma than with the scientists trying to overcome its mysterious behavior. Livermore physicists experimented with a series of mirror machines, just as the Princeton crew built a series of Stellarators, as Los Alamos and Britain investigated the circular "pinch" machines, and as the Russians worked on the mirror, the pinch, and the tokamak. No definitive approach materialized.

Everyone had hoped to catch a glimpse of what they called a "stable" plasma. Instead, all the machines were plagued by "instability," a state of plasma in which any small disturbance in the particle flow amplifies itself until the plasma itself becomes a wobbly mess, breaking through the bonds of the magnetic fields before much of a fusion reaction can occur. An entire vocabulary of instabilities was born – among them the malevolent "flute instability" predicted by Edward Teller, flutelike tongues growing out of the main body of the plasma reaching across the magnetic field lines.

"It was really a pretty depressing time," said Don Grove. He and his friend Wolfgang Stodiek, the West German who had frequented the Princeton exhibit in Geneva, were in charge of Princeton's new C-Stellarator. The machine was the largest and, at $36 million, the most expensive research project the U.S. Atomic Energy Commission – or any other country – had ever undertaken. It was thought that its large size (its racetrack tube was forty feet long) and strong magnetic fields would improve the confinement time of the plasma. It had twice the magnetic strength of the B-Stellarator.

From the very first plasma inside the C-Stellarator, it was clear to Stodiek that the machine would not live up to its promise. An oscilloscope charting the temperature of electrons in the plasma consistently showed genuine fusion activity of no greater duration than a few hundred millionths of a second. For the next nine years Grove and Stodiek perfected experiments to show with certainty that this largest of the Stellarators was having the same trouble as the others, that is, it could not make sustained fusion. The plasma was shuffling out beyond the magnetic field lines at a rate proportional to the temperature. The higher the temperature, the faster the plasma slipped away. High temperature was crucial to the reaction, so the C-Stellarator team was thwarted.

A kind of scientific retrenchment set in. Much of the hard theoretical work of the day focused on predicting and explaining the unstable tendencies of plasma, just as a psycho-analyst might require endless sessions with a patient on the couch to cure a neurosis. But the cure did not emerge.

"It was not clear what the problem was or what to do about it," said Grove.

David Bohm, an American theorist working in the 1950s at Berkeley, had devised a theoretical formulation of the relationship between temperature, magnetic field strength, and the rate at which the plasma slips, or diffuses, away. According to Bohm's equation, even by tightening the magnetic field, the loss of plasma energy in any magnetic fusion device would always be too great for making surplus power. At Princeton, Stodiek and Grove had hoped that the C-Stellarator would finally be able to find a way around the dire predictions of "Bohm diffusion." Instead, at the weekly C-Stellarator planning meetings held in the Princeton conference room, the physicists threw darts at a photograph of Bohm.

The fast rates of Bohm diffusion apparently had also struck Britain's Zeta machine and the Livermore and Russian mirrors, for they could never do better than the Bohm prediction. The situation was bleak not just for Princeton, but for the entire fusion community.

While the experimentalists were throwing darts, a trio of young fusion scientists in California was taking a hard look at the basic assumptions underlying plasma physics, the ideal theory originally laid out by the Spitzer-Teller generation. The chief assumption was that one could think of the superheated plasma as a perfect liquid conductor – that is, an electric current could pass through it without meeting any resistance – and then apply Maxwell's classic laws of electromagnetics.

The California trio consisted of Rosenbluth, then at General Atomic corporation in San Diego, a private research company, and Livermore scientists John Killeen and Harold Furth, the precocious graduate student who had raised alarms about the Zeta results in 1958. By this time, Furth was the newly titled Dr. Furth. He was impatient with the ideal theory that people were using to design the experimental machines.

"People were calculating these wonderful machines, and they turned them on and they didn't work worth a damn," said Furth. "And that is because within those laws of classical physics and electromagnetic theory there's room for all kinds of phenomenon which people just hadn't thought about. They only thought about the smooth equilibrium, but, in fact, the plasmas were capable of all kinds of nasty, turbulent behavior."

Furth and his friend Stirling Colgate had designed a new line of experiments on a pinch machine at Livermore meant to verify whether the ideal theory applied at all. When the experimental results pointed to flaws in the ideal theory, Furth took a step on paper that was quite brazen, even heretical. He revised the ideal fluid theory by dropping the assumption that plasma had perfect electrical conductivity. Instead, he allowed for slight imperfections.

Together with Killeen, a computer expert, and Rosenbluth, an analytic genius, they showed through calculations how even the tiniest amount of resistance to the electrical heating current being pumped into the plasma could grow to disturb the magnetic field lines. The key was envisioning this activity breaking the field lines into magnetic islands between which the plasma could escape. The theory of "finite resistivity" was presented in 1961, and although it began to explain and predict instabilities, it could not correct them.

"We published it, and for a long time it was regarded as a very nice piece of mathematical analysis which might or might not have something to do with the real world," Furth said. Not for almost twenty years would the physicists be able to develop the sophisticated equipment needed to actually view the magnetic islands in detail and put the refined theory to use in running real machines. In the 1960s, the gap between theory and experience remained wide.

Although setbacks plagued the journey toward a reactor, some members of the fusion fraternity, such as Furth, were able to find satisfaction along the way. The goal of a reactor was far off, but there was perfect recreational pleasure to be had in rummaging through the litter of challenging intellectual problems posed by plasma behavior. "Playing in the sandbox" was what critics called it.

Furth was more plasma physicist than fusion zealot, drawn to fusion by the difficulty of its physics rather than by its ultimate end. He could find fulfillment in the solving of discrete pieces of the fusion puzzle. The fact that his problem solving could eventually produce something of historic utility was all the better, but personally, his focus was on the puzzle part at hand.

He was rather matter-of-fact about science and fusion, his remarks devoid of utopianism. He entered the field really by accident, because he happened to be good at working with magnetic fields. The point was to find something intellectually engaging. "The physics is just sort of a medium really," he said. "The pleasure comes from the creative activity."

Plasma physics held his interest for more than twenty-five years, no minor miracle considering Furth was a man who, it was obvious, absolutely detested being bored. He had no patience for dullness and none of Dick Post's appetite for proselytizing the fusion cause to the uninitiated. At physics conventions he roamed the halls in search of interesting dinner partners, unnaturally frightened, it seemed, of being trapped in unstimulating circumstances.

He was a perfect caricature of a brilliant physicist. He was a medium-sized man who carried himself stiffly, his spine straight and his shoulders back. About midcareer he grew a bushy iron-gray beard and wore dark-rimmed glasses. He liked to smoke thin, brown, pungent cigars.

His speech was witty and wry, punctuated by the multisyllabic words one learns for college entrance examinations and rarely uses again. He seemed to have a slight accent, although it was not an accent so much as a sharp enunciation of consonants, particularly at the end of words. Born in Austria, he came to the United States at age eleven. He spoke English with more flair and imagination than a native, and that is what seemed to set his speech apart.

He often dabbled in poetry as a young man, and was even published. One of his more memorable achievements in the field of words was a four-stanza poem poking friendly fun at Edward Teller, father of the hydrogen bomb, eminent theorist, adviser to the Atomic Energy Commission, intellectual leader of the Livermore Lab, and mentor to the fusion clique. Teller had recently pronounced his views on what he called antimatter, a phenomenon that existed in the universe though not on earth. It would explode on contact with ordinary matter, Teller advised.

Furth, who was working for Teller at the time at Livermore, could not resist. He was twenty-six when he filed this poem, "Perils of Modern Living," published in 1956 by the *New Yorker* magazine:

Well up beyond the tropostrata
There is a region stark and stellar
Where, on a streak of anti-matter,
Lived Dr. Edward Anti-Teller.

Remote from Fusion's origin,
He lived unguessed and unawares
With all his anti-kith and kin,
And kept macassars on his chairs.

One morning, idling by the sea,
He spied a tin of monstrous girth
That bore three letters: A.E.C.
Out stepped a visitor from Earth.

Then, shouting gladly o'er the sands,
Met two who in their alien ways
Were like as lentils. Their right hands
Clasped, and the rest was gamma rays.

"When I first wrote it," Furth recalled, "Edward found it very fascinating. He was just flying to Europe with a friend of mine and reading this poem, and he says 'Yah, yah, that's very funny, but tell me, what's a macassar?' He probably looked it up in the dictionary and didn't find the macassar – the antimacassar is the thing you put over the chair to keep the greasy hair off, but you wouldn't find that unless you looked up antimacassar, you see." It was a joke within a joke.

The poem was typical of Furth's wit – mischievousness and irreverence, traits he would carry through his fusion career. The irreverence probably made him a better physicist, but a certain intellectual arrogance came with it, and he could be quite biting when he disagreed with a scientific conclusion or a government policy.

In his approach to fusion research, Furth embodied a central conflict of the field, and as an outspoken person who later achieved great influence, his conflict had implications for others. In the quest for fusion, Furth, like many of his colleagues, had discovered the perfect circumstances to shield his mind from boredom. He was continuously challenged by the elusive plasma, which posed original, elaborate problems with every new twist of the magnetic coils. It was a seductive game, this playing in the sandbox of plasma physics. One could spend a lifetime at it. But fusion research had a deadline, and one had to be careful not to lose track.

As the scientists saw it, there were two divergent ways to attack the problem of building a workable fusion reactor. In simple terms, they were: perfect the basic laws of plasma behavior in theory, then apply these laws, or damn the theory and run experiments full speed ahead until something clicked.

In the period of doubt that followed declassification of fusion research, these two classic approaches to scientific engineering dominated the grinding lab work. Some labs were more theory oriented than others. The scientists were caught between the instinct to understand before acting and the desire to progress quickly toward a concrete goal. The difficult course of fusion research had placed these two urges in conflict. The Princeton director, Lyman Spitzer, for one, was wedded to the theory-before-experiment approach. "You're on much safer ground if you understand it when you're modifying the design," explained Spitzer. "If you don't really understand it, then any little change may upset the effect."

But experimentalists at some labs were game for pushing ahead without total understanding, trying more powerful versions of methods that seemed promising. This was a more empirical method of research. If they could find a way to operate a Stellarator, a tokamak, a mirror, or a pinch machine so that it produced temperatures in the hundreds of millions of degrees, did they really need to know exactly why it worked? One of the experimentalists' favorite analogies was the flush toilet. That invention had been serving humanity quite well for 200 years, yet scientists still could not explain the behavior of water in a vortex.

Neither approach was producing results. The American program went through a "soul-searching period" as Post called it. In more than ten years of work, a heavy technological attack, replete with ever bigger more inventive machines, had not been able to deliver controlled fusion power. Plasma behavior had proved far too complex and disruptive to yield to the simple solution of subjecting it to larger machines or stronger magnets. The physicists called this "brute force," but the basic question of whether a plasma could ever be held stably was unanswered. The scientists were coming to rue their promises of the 1950s. The proof of fusion's feasibility was almost a decade late and still not in sight.

Inevitably, the government patrons of fusion energy grew impatient and uneasy. Budget hearings in Washington became rocky affairs. By 1963, the operating budget for the Atomic Energy Commission's fusion program had reached $24.2 million, not an insubstantial sum at the time. Was it worth it? As an example of the negative attitude that was developing toward fusion research, the physicists often dredge up a quote from Senator John Pastore of Rhode Island, who said at budget hearings in the spring of 1964: "I am wondering in my own mind how long do you have to beat a dead horse over the head to know that he is dead?"

Pastore went on, "Is this not indeed a very expensive way of getting this basic knowledge? We can build these machines until the cows come home. Somewhere along the line somebody has to think that this is a lot of money and maybe we ought to be putting it into some other place where it may be more productive."

Not only Congress was asking questions. Some scientists were also unhappy. "We were in a state of dissatisfaction and ferment, feeling that the program was taking the wrong course," said Furth, who was perennially offering West Coast dissidence. Too much "junk" machinery was being built without "enough enlightenment," he said. Post and Furth together wrote a paper calling for a "new orientation" for the controlled fusion program. It essentially sounded a retreat from the ambitious reactor goal into basic plasma physics.

The scientists first needed to discover the laws of making a stable plasma, they argued, irrespective of creating fusion. That would come later. Small research devices designed specifically for studying plasma, as opposed to fusion, made good sense economically and scientifically. "Whether any of the tools of research that will be used could be pictured as reactor prototypes is of secondary importance," they wrote.[1] "What is needed is a greater recognition that physics rather than technology is the present avenue to success in finding the ultimate solution of the fusion problem."

In their arguments for a basic science strategy to fusion development, Post and Furth used an analogy showing the perils of prematurely choosing a specific technological route to a goal. Choosing too soon could lead to dead-ends.

"This is the one about the primitive culture that decided it was going to go to the moon," recalled Post, "and they had invented the balloon. The idea was that they would just build better and better balloons and obviously each balloon went higher than the last and if we just keep in this formation our balloon will get us to the moon."

The instinct – and the pleasure – of retreating into basic plasma physics was not just an American trend. It was also taking hold at Britain's Culham Laboratory. "There was a period from 1960 to say 1967 when we hardly ever spoke about fusion," recalled Roy Bickerton. "We were doing all this interesting basic physics." In 1966, a government review committee looked into the situation at Culham and, said Bickerton, came away "appalled by the fact that 99 percent of the people in Culham were not even thinking about fusion power. All the effort was concentrated on this type of physics." The outside review in Britain triggered a 50 percent cut of Culham's already meager annual budget of £4 million.

The crux of the matter was that, over the years, the scientists had come to regard themselves not as fusion researchers but as "plasma physicists." After such high hopes and so much failure, a conservatism had gripped the community. Rightly or wrongly, it started to pull back from its original, concrete goal of a commercial fusion reactor into the more rarified world of atomic physics.

It was during this dispirited period of the mid-1960s that Lyman Spitzer, the founder of American fusion research, decided to quit the field. Looking back over their own history, the fusion physicists find this fact particularly significant. Many physicists who did not know Spitzer, especially the younger ones, assume he left in 1966 because he had given up on fusion. It is symptomatic of the new generation's ambivalence that its members would so easily reach this conclusion. According to Spitzer and to people who worked with him, the reasons behind his departure were more personal.

As the Princeton lab and the AEC national program had grown post-Geneva, Spitzer became more deeply involved in administrative matters. Gradually he handed them over to Gottlieb, who was made director of the lab, while Spitzer became chairman of the executive committee and continued to oversee scientific research. Then, in 1966, Princeton University offered Spitzer the chairmanship of the university's general research board. That would have meant three jobs for Spitzer – scientific head of the plasma physics lab, chairman of the astronomy department, and chairman of the research board. "It was clear I couldn't do all three at one time," Spitzer said. "I chose to drop out of the plasma lab."

This is the simple explanation, but underlying the decision was the fact that, for Spitzer, his career and interests had always encompassed more than the fusion mission. He was not as single-minded a crusader as California's Dick Post. Fusion was supposed to have been a ten-year commitment, not a twenty-year puzzle still unsolved. Spitzer had left behind his beloved astrophysics, the study of the origins of stars, in order to pursue a detour that had grown out of Cold War bomb work. It was time for the astrophysicist to return to his own road.

Beyond the lure of his other interests, however, was the surprising fact that Spitzer, at age fifty-two, felt a bit outmoded in the modern fusion field.

"I felt I'd made most of the contributions that I could make," he said during an interview in his office in Princeton's astrophysics department. "My method of research was more useful in the early days of plasma physics than it had been in later years. The subject has become more mathematical, becoming more and more difficult to follow in one's mind what the plasma is doing. It's more and more a matter of getting detailed, rather complicated equations," he continued. "[It's] a combination of applied mathematics and numerical computations, and I can't do such things."

Gottlieb seconded that version of events. "He recognized he had unique capabilities at the beginning, and now a new generation was ready to take over," Gottlieb said. "He wanted to go over to his first love. The requirement for Spitzer no longer existed."

Whatever the reasons, Spitzer's departure from fusion dampened many spirits in the field. "It was a tremendous loss to us. As far as we were concerned it was the end of the world," the theorist Frieman said.

It was, at least, the end of an era. The promise of fusion offered in 1951 had not been delivered, and there remained no hope for making it happen any time soon. The Stellarator had failed. The pinch machine and tokamak were unstable. The mirrors were leaky. In terms of fusion's history, the scientists of the first decade were indeed a primitive culture, releasing trial balloons to the moon.

5
Dawn of the tokamak

Lev Artsimovich was quick to criticize his colleagues in the West. But as they all knew, his team at the Kurchatov Institute was mired in the same quandary that had produced such sluggish results at Princeton, at Britain's Culham lab, at Livermore, and elsewhere.

To survive, fusion desperately needed a breakthrough – not the false hope of Zeta, not the procedural euphoria of Geneva, but a genuine scientific turning point.

In 1968, the Russians delivered it in a package called "tokamak." Yet the fusion community had grown so darkly skeptical that a year's worth of dickering and data collection would pass before the scientists were convinced that they had truly taken a leap forward.

Of all the plasma devices flashing and rumbling in the world's fusion laboratories, there was only one that, after so many years of open research, remained unique to one country. This was the tokamak, and it retained the Russian imprint largely because the fusion community outside the Soviet Union simply did not think much of it. Of course, the Russians had mirror machines, and they were routinely comparing results with the mirror experts at Livermore. In Moscow, however, the mirror experiments were conducted by a subordinate of Artsimovich. The Russian boxer who was director of fusion experiments at the Kurchatov lab preferred to nurse along several versions of the machine that had been invented in 1951 by Sakharov and Tamm, the one that bore the acronym tokamak, for "toroidalnya kamera ee magnetnaya katushka," the toroidal chamber and magnetic coil.

Artsimovich had worked for fifteen years on the tokamak. He maintained a firm belief in the system, just as firm as his disdain for other approaches.

The tokamak looked like a fat, wired-up, steel doughnut. The underlying principle was the same as the "pinch" concept devised in Britain. When an electric current passes through a plasma cloud, a magnetic field is created around the plasma. This field "pinches" the plasma into its own invisible, magnetic bottle. The tokamak had the added feature of magnetic coils belted vertically around the outside of the chamber to add a second, stabilizing field. All the coils made the tokamak a complex, cumbersome machine from an engineering point of view.

By contrast, the Stellarator plasmas had no self-generated magnetic field. The containing fields were produced primarily by coils wrapped in a spiral or helix around the exterior of the racetrack-shaped machine. The mirror machines also relied on exterior magnetic coils.

Outside of their home labs, the pinches and tokamaks were generally considered poor devices on which to run meaningful experiments. It was simply too difficult to explore their heating and plasma-confining properties separately since both properties arose from the same strong electric current shooting through the plasma. The scientists could not turn up the heating without affecting the magnetic confinement.

And there was another problem. The current required for a self-pinching plasma was so power draining that it could not be sustained indefinitely by modern electrical equipment. Instead, the current had to be delivered to the plasma in short pulses. With the plasma being heated and cooled in pulses, any energy that might result from the reaction would itself follow a pulsing, on–off cycle. This alternate heating and cooling, it was thought, would make the tokamak quite unsuitable as a power reactor. A power-smoothing mechanism would be expensive and the frequent heating and cooling of metal parts would lead to mechanical fatigue in the machine. In 1951, Lyman Spitzer had considered and then rejected a tokamak configuration for his experiments because he felt it would ultimately be impractical as a power plant.

The first cousins of plasma-heating devices were the pinches and the tokamaks, and the familial relationship led to a natural alliance between the British and the Russians. They seemed to be the only fusion experimenters still interested in learning how to confine a current-carrying plasma. In turn, the scientific alliance fostered a friendship between Lev Artsimovich and Sebastian Pease, a politically as well as scientifically adept British physicist.

As a young man, Pease had been one of the experimentalists pictured in the British newspapers grinning alongside the infamous Zeta pinch machine. Pease still felt a measure of chagrin over the Zeta debacle, the way unverified results had been prematurely presented to the world's press, which promptly elevated the news to historic proportions. "The Mighty Zeta" the headlines had said, "Limitless Fuel for Millions of Years." And then the embarrassment of a retraction months later.

Although the fateful Zeta announcement had not been of his making – that was handled by the higher-ups – Pease felt a measure of personal responsibility for repairing the reputation of Zeta and fusion science.

"It seemed to me that it was our job to put the physics right and on the record and do a decent job," he recalled years later in an interview. "It was a real physical challenge which it seemed to me we shouldn't walk away from. Quite a lot of my colleagues thought we were mad – 'that thing will never work; you're talking rubbish, Pease' – and not only will it never work, but you could never understand it."

Through the 1960s, Pease continued to work on Zeta, searching for ways to make a current-carrying plasma stable. Pease and Artsimovich arrived at an informal arrangement under which they divided the research on their related machines and shared their results, determined to save time and money – and perhaps their science.

"Lev said he would do the tokamaks, and we said we'd do the rest," Pease explained. "'Don't let's duplicate each other's apparatuses,' was the sort of line he was taking. Of course it helped him because he thought he was onto the winning line, but nonetheless we were suffering very severe budget cuts."

Artsimovich, it seemed, was indeed onto the winning line. By the mid-1960s, he began to see strong results from the tokamak. Temperature readings had improved and so had confinement times – the fraction of a second that the plasma was held stable before escaping its invisible bottle and cooling. The depressing theory called "Bohm diffusion" had predicted that plasma would never remain stable long enough at high temperatures to produce surplus energy. But now the

Kurchatov team was seeing readings that surpassed the time limits on plasma life predicted by Bohm's equations.

Artsimovich's first glowing reports to his worldwide colleagues were met with harsh skepticism. The Americans, and especially the Stellarator defenders at Princeton University, looked particularly askance at the Russian numbers. At best, wrote Lyman Spitzer, allowing for experimental error, these results were commensurate with the Stellarator's performance. In the meantime, Pease's researchers in Britain had also happened on a more stable modification of the Zeta pinch, although the confinement times were still too fleeting to offer much hope. The British scientists delivered some credence to Artsimovich after calculating that, at least in theory, a more stable version of the tokamak should also be possible.

It was not until 1968, a full ten years after worldwide declassification of fusion research, that Artsimovich announced with some relish the arrival of the age of the tokamak. Despite what the Western scientists thought, the Russian now provided striking numbers showing that the tokamak was capable of creating what real fusion required, a hotter and more stable plasma.

The setting for the Russian announcement was the triennial conference of the International Atomic Energy Agency, the "Plasma Olympics," being held for the first time on Soviet soil. The scientists convened in Novosibirsk, the new, elite academic community tucked away in Siberia. The whole cast of international fusion players was on hand – the Britons Pease, Roy Bickerton, and Alan Ware; the Americans Harold Furth, Marshall Rosenbluth, Dick Post, Mel Gottlieb, Don Grove, and hundreds more from Britain, the United States and elsewhere.

It was a lovely, unusually warm August, and the scientists were reminded of Mediterranean climes as they bathed at the lakeside beaches between lectures. The conference hall was surrounded by a fragrant pine woods. "You didn't feel you were in Russia when you were in Novosibirsk," said Hugh Bodin, one of Pease's scientists.

The Russians were obviously intent on putting on a good show for these learned international visitors, and Artsimovich's news was the headliner act. The pugnacious experimentalist stood before his peers and made the spectacular declaration that two Kurchatov Institute machines had conquered Bohm diffusion. The T-3 tokamak and the TM-3 tokamak, the latest versions of the Sakharov-Tamm invention, had produced electron temperatures of more than 10 million degrees centigrade and typically held onto the plasma for 10 thousandths of a second, about fifty times better than the confinement time predicted by Bohm's formula for plasmas of that temperature. In some instances the confinement was as long as 20 thousandths of a second.

By contrast, Princeton's C-Stellarator typically produced plasmas of just under 1 million degrees which lasted about one thousandth of a second before escaping the magnetic field and dying against the cool vessel walls.

Artsimovich lived up to the reputation he had acquired after Geneva. With an air of self-assurance, he reminded his audience that, at the 1965 world conference, he had first hinted that the tokamaks could surpass Bohm's formula. Then he added, "It was received with remarkable suspicion. At the present time, nobody is any longer surprised at this behavior since the belief in the universal character of Bohm's formula is completely shaken."[1]

The key to the tokamak's success was its fat, squat shape. For the same volume of gas, it allowed for wider plasmas and a shorter trip around the machine than the racetrack-shaped Stellarator. This created a tighter, more symmetric confinement. The latest tokamak experiments had finally been able to demonstrate the machine's potential, in part, because of improved cleaning techniques for the vacuum vessel walls that reduced heat-stealing impurities. Also, the magnetic field alignment was more precise. The importance of an exact geometry had not been fully appreciated at first.

At the end of his talk, Artsimovich revealed plans for an even larger tokamak that could produce reactor-sized plasmas with confinement times of tenths of seconds, a near eternity in plasma physics. If it could be done, the fusion from such a lengthy reaction would provide the long-awaited proof of scientific feasibility.

After a quiet retreat from Novosibirsk, the international community of scientists reacted to the news of the tokamak achievement with amazement and reflexive skepticism. They had witnessed the Peron hoax, the Zeta disaster, and fifteen years of frustration with Stellarators, pinches, mirrors, and the rest of the experimental machines.

Ironically, the tables were now turned on Artsimovich. Peer review was a two-way street. Just as he had savaged Fred Coensgen and his Livermore colleagues for blundering with their mirror machine results in 1961, Artsimovich was now being subjected to vigorous questioning in the open fusion theater. Among the biggest naysayers were the scientists at the Princeton lab, especially Furth and the West German, Wolfgang Stodiek. After many years at Livermore, Furth had been wooed to a position as head of the experimental division at Princeton, bringing his dissenting instincts with him. He joined forces against Artsimovich with the skeptical Stodiek, who had now spent eight years working on the balky C-Stellarator at Princeton.

The one-two punch of Furth and Stodiek was formidable. How had the Russians arrived at their temperature measurements, the two scientists wanted to know? Artsimovich and his physicists had taken only indirect measurements of the plasma. They were able to measure the outward pressure of the plasma cloud, then, assuming the plasma temperature was uniform, they had made elaborate calculations to infer the temperature. Their readings during multiple plasma attempts on T-3 and TM-3 ranged from a low of 1 million to a high of more than 20 million degrees, a variation in performance they could not explain. Plasmas for typical densities had settled at a median 10 million degree mark. The imperious Princeton critics took the same indirect data and produced a different interpretation. Stodiek wrote a forceful analysis attributing the high tokamak temperatures to misplaced readings on runaway electrons, a select group of speeding particles whose high temperature did not represent the overall temperature of the plasma itself. This had been the mistake that undid the Zeta team.

Such was the primitive state of plasma diagnosis and the tentative state of plasma physics theory: two great laboratories, Princeton and Kurchatov, could disagree so mightily over the same data. Were it not for the sense of shared destiny that the physicists had forged, the controversy might have been left at that. In the new age of cooperation, a defeat for one lab was a defeat for them all. And success could bring good fortune to everyone. It was not merely an abstract feeling of connection and responsibility. Artsimovich was a known personality, a colleague, a respected intellect, and a man with a reputation for caution.

For years they had joined in toasting the prospect for quiescent plasmas at conventions all over the globe. He had shown them around his lab and visited some of theirs.

This was not an anonymous Soviet official making an outrageous chauvinist claim. This was Artsimovich, cynical and rigorous. The report from Novosibirsk could not be summarily dismissed.

Still, the fusion fraternity needed proof of Artsimovich's fantastic numbers. He must have sensed this, for he had already laid the groundwork at Novosibirsk for an independent scientific review of the Kurchatov lab's work.

During the conference in Siberia, Artsimovich invited some British colleagues to a meeting at his nearby dacha. Sebastian Pease was among them. Like many of the Britons, Pease had been more willing than the Americans to accept the possibility that the tokamak was producing real results. Pease had even begun to wonder if it were not time for Britain to build a tokamak of its own. Zeta seemed at a dead end. The pinch machine had reached temperatures of 1 million degrees but the confinement time was terrible. On the other hand, the Soviet Union's tokamak results were uncertain. Pease asked his Russian friend for advice. He recalled that Artsimovich advised against building a British tokamak, because the Russians did not yet fully understand how their device worked.

Then Artsimovich presented a suggestion. Aware that Culham Laboratory scientists had just perfected a revolutionary device for detecting plasma temperatures, he invited the British to visit Moscow to measure the temperature inside the T-3 tokamak. If the new measuring device could be attached to the Russian machine, the world might know, unequivocally, what the Russians had achieved.

The new British thermometer used the light of a laser. Unlike invasive probes, the laser could be directed into the hot plasma without disturbing its movement. The light beam would scatter off the plasma particles in proportion to the temperature of the plasma. The spectrum of light emerging from the window of a test reactor could then be "read" by light-sensitive devices outside the reactor. By measuring the spectral light coming from the window, scientists could calculate the temperature and density of the particles inside. Although the world's physicists had been investigating plasma for nearly two decades, the laser scattering device was their first direct and reliable method for measuring plasma temperature.

Pease, who had risen to director of the beleaguered Culham lab in 1967, was eager to help Artsimovich. The Russian had been touting the tokamak for years. If the latest T-3 results from Moscow could be verified, it would go a long way toward rehabilitating the entire fusion community. The existence of a viable fusion machine making long-lived, hot plasmas would be a tremendous boost. Not only would it save the Soviet Union's tokamak program, it would also enhance Britain's vulnerable fusion program, the laser scattering technique, and possibly Britain's international scientific reputation.

In the dacha at Novosibirsk, over a table set with wine and fruit, the Russian and the Briton agreed to an historic collaboration. There were no precedents. Although physicists had visited labs in other nations and ideas had been exchanged, nothing of the scope of this joint, East-West endeavor had occurred. The Kurchatov Institute was, after all, a classified weapons research lab ordinarily closed to foreigners. Experience with the British laser device would surely constitute a high-technology windfall for the Russians. It was a bit of a gamble, but if the two men could convince their governments to allow it, the British would send a scientific team to Moscow to attach the laser scattering device to the T-3 tokamak and judge for all the world whether Artsimovich and the Kurchatov Institute physicists had achieved a monumental success or hatched another of fusion's embarrassing failures.

Mindful of the obvious risks, government officials in Great Britain and the Soviet Union saw prestige to be gained on both sides, and they eventually agreed to the bold Artsimovich-Pease collaboration.

International political events threw up a temporary roadblock, however. In August 1968, just days after the end of the Novosibirsk conference, Soviet tanks crossed the border into Czechoslovakia to put down an incipient revolt. Under the circumstances, the British Atomic Energy Authority could not grant Pease wholehearted approval for the trip. Pease recalled that the scientists were not actually forbidden to go to Moscow, but neither were they given official approval. It was made known that the research would have to be an independent venture, not sanctioned by the British government.

A decent diplomatic interval of six months passed. In February 1969, a chartered airliner left England bound for Moscow. It carried a three-man team of British scientists, led by Nic Peacock, and five tons of equipment, including a spare computer. Awaiting the team's arrival at Moscow's Kurchatov Institute was a thin, rapid-speaking young British scientist who had been sent ahead to organize the project. Although Derek Robinson, then twenty-seven years old, had been married barely a few months, he and his new bride, Marion, readily agreed to a stay of perhaps a year in Moscow. It would be, they thought, an adventure.

They would not be disappointed. Robinson was intensely curious about the Russians. He was a very able and facile scientist, adept at both theory and experiment. He had learned his trade on the aging Zeta, mastering the engineering and physics of the new laser scattering device. Most recently he had fashioned a mathematical proof for Pease showing that some version of the tokamak concept should create stable plasmas. Having performed that task, he was eager to visit the wellspring of tokamak research and probe the minds and machines of Soviet science.

Arriving in November 1968, Robinson felt his reception at Kurchatov was at first rather cold. The scientists were uncertain about how to act, and there was a clear defensiveness that, Robinson surmised, might have had something to do with Western criticism of the invasion of Czechoslovakia. The nature of Robinson's mission – to ratify or discredit the Russians' findings – might have contributed to the chill he encountered. The tokamak researchers were quite confident of their results. They were neither as enthusiastic over the British laser project as their leader, Artsimovich, nor as intent on gaining Western approval.

Under the rules of the exchange, Derek and Marion Robinson were to live as the Russian scientists and their families lived. They stayed in the same kind of apartment house, they were paid in rubles, and they bought their groceries at the local market. It was something less than the adventure they had had in mind. Marion Robinson spent each day standing on long lines with other Muscovites in order to assure that food would be on the table. Derek Robinson recalled that sometimes his wife would return home from these food expeditions emptyhanded and crying, having failed to comprehend the subtle rules of Russian queuing and place holding.

After just two months in Moscow, cut off from news of the West, anxiously awaiting his British colleagues, hearing rumors in the Kurchatov lab of battles on the Chinese border, Derek Robinson suddenly took sick. The illness sparked the first political dispute of the exchange. According to the Russian doctors' diagnosis, Robinson had an ulcer, and, judging the matter serious, they wanted to hospitalize him immediately. Robinson consulted the British embassy doctors, who were less alarmed. They told him not to have anything to do with Russian medical treatment and that the problem could be handled with drugs and a careful diet that included more milk.

For a moment, at least, the future of fusion energy seemed frozen in an East–West medical standoff. The Russians, said Robinson, "considered that I was putting my life in jeopardy, and they wanted me to sign a paper saying that they were absolved of all liability. I spent a long time trying to assure them that in the opinion of our medical people this was not the case, and that really it was not unusual in a high-stress situation of that nature for this to arise – and that, really, with suitable rest and diet and so forth this could be kept under control." Eventually the Russians were persuaded and no paper of absolution was required. Within a few weeks Robinson was feeling well enough to return to work.

Finally in February, Robinson's colleague Peacock and his men arrived with the equipment. All experimentation on the T-3 tokamak had stopped to accommodate the visitors' mission. Under Robinson's direction, the Russians had cut a window into the tokamak vessel to house the laser device, and the British scientists set to work connecting it. The project went more slowly than planned. Improving the hookup was difficult because parts were not readily available. There were unexpected problems, too, with the laser's timing and power supply. This was not a simple valve job on a roadster. The technology demanded exact tolerances, precision, and perfection.

"By the time it got toward June," Robinson recalled, "we actually hadn't gotten any results. We had all sorts of problems. The Russians were for throwing us out and were saying 'Go away because you're wasting too much of our time.' Artsimovich came himself and said, 'You've really got to get some results before this date or we're going to take it off.'"

With Artsimovich's threat ringing in their ears, the Peacock trio returned to Britain to press their advanced computers into service on a solution. Robinson remained behind to complete the hookup, awaiting instructions by phone and drinking plenty of milk.

In the spring of 1969, while the Britons pored over data, Artsimovich visited the Massachusetts Institute of Technology for what must have been, at times, a frustrating ordeal. He was encouraged to deliver several lectures on tokamaks, and he also conducted private sessions on more technical details. Teams of physicists from other American labs made the journey to MIT for an audience with the tokamak guru. Now, Artsimovich was reporting updated figures of 100 times better confinement than that predicted in Bohm's diffusion theory. These enthusiastic reports, however, were outwardly received with the same mix of curiosity and measured skepticism as before. Bruno Coppi, an Italian physicist at MIT, recalled that after one lecture Artsimovich retreated to Coppi's office and complained, with tears in his eyes, that people just did not seem to be listening to him.

In truth, the entire American fusion program was hanging on Artsimovich's every word, waiting and watching and listening. The tokamak claims were sounding more and more interesting, particularly in light of the unending troubles at Princeton with the C-Stellarator. If the Russians were correct, the tenets of the fusion faith might be changed forever.

Inspired by Artsimovich's lectures, scientists at MIT began to consider putting their acclaimed expertise with magnets to use on a tokamak. Until that time they had contributed only theoretical work to the American fusion effort, staying out of the hardware competition. Bruno Coppi was assigned to work up a tokamak design. Even while calculations by Furth and Stodiek at Princeton were yielding a negative report on the tokamak results, Amasa Bishop, head of the U.S. Atomic Energy Commission's controlled fusion branch, was preparing to go before Congress with the tokamak news – if it could be verified – and ask for a doubling in funding of the U.S. program. It was unthinkable that the Russians could be permitted to gain an advantage over America on such fundamental research and to be first to claim the proof of scientific feasibility.

All the while, the Princeton lab director, Mel Gottlieb, was fielding probing questions concerning the American program and the Russian tokamak from "people on the outside who didn't understand these claims. We kept on saying, 'We're waiting for verification.' The most important question is, what is the electron temperature? They do not have a direct measurement of the electron temperature, and they've got to make that. That will decide the machine one way or the other."

Gottlieb began to consider the possibility that the United States should build a tokamak of its own merely to answer the question of its worthiness once and for all. He went so far as to ask some of his physicists to assess the difficulties of converting the C-Stellarator into a tokamak.

An increasingly nervous Atomic Energy Commission called a meeting in June in Albuquerque, New Mexico, of the top fusion leaders. The aim was to evaluate the purported Russian breakthrough and to decide how the United States should respond. The commission had asked Fred Ribe of the Los Alamos national lab for a comprehensive report on the Russians' T-3 results. Focusing on one portion of their data, Ribe argued that the T-3 results were valid and impressive. The Princeton people – Stodiek, Grove, and Furth – concentrating on another set of results, made the opposite case, saying the Russians were being fooled by the phenomenon of runaway electrons.

"They were both convincing," Gottlieb recalled, "but no interpretation fit all the data. This had gone on too long. I said, 'It's too important to wait any longer. We've got to find out for ourselves.'"

The AEC commissioners, tending to believe the Soviet claims, had been pressuring Princeton to convert the C-Stellarator to a tokamak as the quickest and cheapest way to make a U.S. tokamak. Princeton had estimated the conversion costs to be under 1 million dollars. By the end of the Albuquerque meeting, Gottlieb had come to accept the commissioners' strategy. At one of the noontime breaks, he took a swim in the hotel pool and met his deputy, Furth.

"I said, 'The decision is made. We're going to make a tokamak.'

"And to my surprise – he had been very much against it – he responded, 'Well, maybe it's the right thing to do.'"

Thus in a half-hearted way Princeton turned to tokamaks – not because Gottlieb and Furth believed in the concept but rather for the purpose of showing that the Russian machine could not work. For the AEC it was a prudent hedge against Soviet success. Furth laid a fistful of five dollar bets with his colleagues that Robinson and company would measure less than half the temperature Artsimovich claimed. And he expected that a Princeton tokamak would also unmask the sad truth. Amid great unhappiness at the lab, Gottlieb ordered his men to start closing down the venerable C-Stellarator in favor of the disputed Russian invention.

The British laser team – and especially Derek Robinson – felt a growing weight of international expectation. Everyone seemed to have gotten hold of the telephone number for Robinson's Moscow apartment, and the young physicist heard regularly from his bosses in Britain and even from the American Amasa Bishop, then director of the U.S. fusion program. Peacock and his crew shuttled back and forth to Moscow and relayed design suggestions from England. Robinson improvised modifications on the machinery, pushing hard until midnight many nights.

The Russians now were spectators, watching this curious process with an increasing sense of impatience and an occasional sense of humor.

Vladimir Muchovatov breaks into a sly grin when he retells the story of the fake letter from Princeton, a letter that arrived on Artsimovich's desk before any hint had surfaced of the American plan to transform the C-Stellarator into a tokamak.

Muchovatov, a physicist just a bit older than Robinson, and three of his Russian friends at Kurchatov decided to play a practical joke on their boss. They composed a fictitious letter to Artsimovich from Mel Gottlieb, the Princeton lab director. To be sure the letter was idiomatically correct, they took a draft to Robinson and asked if he would touch up the English. Robinson went along with the gag. One of the Russian foursome who had visited Princeton happened to have some of the

American lab's letterhead stationery. Using old correspondence from Gottlieb as a guide, they held the stationery to the window and traced the American's signature beneath the typed note.

In the fabricated letter, Gottlieb said that Princeton was dissatisfied with its Stellarator program and had decided to cease operations on the machines immediately. As an alternative, Gottlieb was proposing to buy from Artsimovich his T-3 tokamak. Would Artsimovich please respond?

Muchovatov and company, unaware of the latest developments in Albuquerque, obviously thought such a proposal would be a preposterous turn of events, given current world opinion of the tokamak, particularly at naysaying Princeton.

The conspirators folded the letter and placed it in a Princeton envelope, then quietly left it one evening on the desk of Artsimovich's secretary. The next morning, before the usual 10 a.m. organizational meeting, Muchovatov got a call from Artsimovich.

"Don't start the meeting without me," his boss warned. "I have some great news."

About thirty scientists were gathered around the conference table when Artsimovich arrived in fine spirits. Now the four conspirators began to worry. They never thought it would go this far. Surely he should have seen what an absurd letter it was. Muchovatov realized that Artsimovich was about to really embarrass himself, and he tried to dissuade his boss from making the announcement in front of the group.

"Why not keep your news until after the meeting?" Muchovatov suggested. But Artsimovich insisted.

"Princeton is ending its Stellarator program," he announced, "and Mel Gottlieb has asked to buy our tokamak."

Around the room there was some stifled laughter. When the meeting broke up, Muchovatov took Artsimovich aside.

"Lev Andreevich," he said, "the letter was from us."

"No, impossible. I know Gottlieb's signature."

"Lev Andreevich, we traced it at the window."

"But this is good English! You could not possibly have written this."

"We had Derek Robinson go over it," Muchovatov admitted.

Eventually Artsimovich was persuaded that the letter was a joke.

Though normally a skeptical person, the tokamak builder had wanted so much to believe that the world had accepted his success that his vision was clouded.

Muchovatov braced for a reprimand, but Artsimovich took the joke well. He should have seen that such a letter was ridiculous. Princeton, the world's Stellarator experts, would never beseech him for a tokamak. Of course not.

By August 1969, nearly a year after he arrived in Moscow, Derek Robinson had the long-awaited T-3 results in hand. According to the British laser scattering device, the electron temperature measurements the Russians had announced in Novosibirsk were correct.

The T-3 tokamak had performed superbly, concocting plasmas that were typically 10 million degrees centigrade, and there was no indication of the runaway electrons that Furth and Stodiek had predicted. Moreover, the confinement time was excellent, up to 20 thousandths of a second, just as the Russians had claimed.

The Bohm diffusion theory had been conquered. Still, no one knew precisely how the plasma behaved inside the tokamak, but that might be answered another day. For now it was enough to know that the tokamak's fat shape and strong magnets satisfied the demands of the plasma. A device had been found that might eventually make power-producing fusion. The tokamak worked. That is what everyone had been waiting to hear.

Robinson telephoned Culham lab in England with the ground-breaking news. Sebastian Pease was ecstatic. The British relayed the results to the U.S. Atomic Energy Commission in Washington. The news was so momentous and the reaction so emotional that one Washington staff member leapt onto a table for a joyful dance.

For years they had been surrounded by disappointment and failure, confused about where to turn next. Now, suddenly, the Russians – and the British – were showing the international fusion community a way out. The tokamak verification was the first big payoff of cooperation in fusion research, and it touched off an international stampede into tokamak technology. Over the next five years tokamaks would sprout across Europe, Japan, and the United States, beginning with Princeton, which had already started dismantling its C-Stellarator when the British ratification came. MIT would enter the competition of machinery with the small tokamak designed by the Italian Bruno Coppi.

Erasing its record of negativism, Princeton embraced the Russian machine with enthusiasm. The quick change of heart miffed some international colleagues. Said the Briton Hugh Bodin: "They converted their C-Stellarator into a tokamak almost overnight with a shrug of the shoulders and without so much as a twinge of conscience and said it was great."

After nine years of work on the C-Stellarator, Don Grove supervised its disassembly and metamorphosis. He had no regrets, he said. He was just eager to make progress, and it looked now as if the tokamak was the ticket. To convert the Stellarator into a circular tokamak, the straight sides of the racetrack were removed and the U-shaped ends joined. In just six months the job was completed, and the new Princeton tokamak had duplicated the results of the Russian T-3.

A short while later Lev Artsimovich paid a visit to Princeton, and Grove proudly escorted him to America's first tokamak. He recalled how Artsimovich looked it over, obviously impressed, and then said to Grove, "I think you had better stick to Stellarators."

The British laser scattering measurement on the T-3 tokamak marked the beginning of the modern age of fusion research. It was a giant leap from temperatures of a few million degrees on the old machines to 10 million degrees on the tokamak. It was a plateau from which one could imagine launching an eventual second leap into the 100-million-degree regime of fusion reactor plasma. Moreover, another important advance had been made. The scientists no longer needed to run their experiments blindly. They had the tools to measure the plasma with confidence. The dream of a fusion future was restored.

6
Building big science

In their balloon-to-the-moon treatise, Harold Furth and Dick Post had counseled the fusion community to stop building expensive test reactors and return to the oracle of basic physics. That advice was drowned in the wake of Artsimovich's tokamak triumph.

"A very wrongheaded paper," said a chastened Furth, in retrospect. "The Russians never learned the basics. No one yet understands how tokamaks work. This proved no impediment to them getting on with the job."

In the 1970s, the job of fusion gradually grew to proportions Furth had only envisioned in the hazy, far-off future. A benevolent new contagion spread through the fraternity. It was tokamak fever. Not only did most of the scientists eventually succumb, but government leaders in the energy field also found the promise of the tokamak supremely alluring. Using complex computer models and an array of chalkboard equations, they came to understand a new dimension in the pursuit of fusion energy. For many, the Russians' success was an invitation to drop the ballast of marginal fusion devices and design a bigger and better tokamak.

Furth's hindsight was utterly correct. The Russians had succeeded not by dissecting the how's and why's of tokamak physics but primarily by empirical scientific methods – by taking cues from the plasma's behavior as observed in experiments. They had simply pursued what seemed to work without stopping to nail down precisely why it worked. In fact, many of the strange effects occurring inside the tokamak were filed under the fancy label "anomalous," meaning only that they were odd and could not be explained.

The Russians were not about to pause for profound understanding. In the tokamak, Lev Artsimovich had discovered "the winning line," as his friend Sebastian Pease put it, and the Russians were determined to pursue that line to the finish. Their aim was to stoke up the plasma furnace. Following the British seal of approval for Moscow's T-3 machine, Soviet scientists openly discussed ambitious plans. They set out to design a series of increasingly larger tokamaks, culminating in a machine called T-20 that could be finished within the decade. Fantastically high temperatures would be achieved and with these temperatures superior confinement of the plasma, they theorized. If T-20 worked, the Russians would at last be able to prove the scientific feasibility of fusion power.

Their confidence in larger tokamaks was built upon a simple application of physics. "Scaling laws," they were called. If the plasma cloud is expanded inside an enlarged magnetic bottle, colliding particles will take longer to find their way to the cool edge. The longer the plasma exists, the greater the chance that enough particles will collide to achieve a continuous fusion reaction. The physicists believed that changes in scale could reap big gains. Increase the plasma's radius by, say, a factor of two, and one would derive at least a fourfold improvement in confinement time. In fusion, bigger was definitely better.

As it had before, the West reacted to the Soviet ambition with alarm and a reawakened ambition of its own. No sooner had the first generation of T-3-sized tokamaks sprouted in Western labs than the researchers coveted an even larger variety. The "scaling laws" were immensely persuasive, but equally immense sums of money would be needed to take the leap into fusion's future. A machine in the United States fit to outdo the Russians' immediate successor to T-3, the T-10, could cost several tens of millions of dollars, the designers estimated. To build and operate a machine of that size and complexity would required hundreds of people: a large team of physicists backed by squads of engineers, computer experts, financial administrators, and support staff. In short, it would require a bureaucracy. The new machine would gulp huge amounts of electric power to ignite each test plasma. Operation and maintenance costs would enter a realm that no fusion laboratory had ever considered.

The 700-ton, 30-foot-tall, Tokamak Fusion Test Reactor (TFTR) at Princeton, N. J. Courtesy PPPL.

"The machine doesn't have the perfection and grace that plasma has."—ROB GOLDSTON, Princeton physicist.

Before underwriting such a venture, the national governments would insist on increased participation in lab decisions. Accountability would be demanded. A new layer of oversight teams and middle managers would be needed. It did not take a genius to see that the government's commitment would ultimately change the character of the research. Fusion would no longer be the intimate purview of a handful of fraternal scientists. To take on the Russian challenge, American, European and, eventually, Japanese scientists were inviting a transformation of the fusion laboratory and the cloistered working life they had known.

On three continents, fusion was about to become Big Science.

America's fusion labs did not exactly leap in unison as they stood on the verge of the big-budget era, however. Each needed a shove, and the chief muscle man was Bob Hirsch.

Aggressive, young, public-relations minded, Robert Louis Hirsch was a senior staff physicist at the U.S. Atomic Energy Commission's Division of Controlled Thermonuclear Research in 1970. He was thirty-five years old and raring to put fusion onto the fast track. The Princeton lab director, Mel Gottlieb, derided him as "a young man in a hurry" and he was. Controlled-fusion research was then still a tiny $30 million corner of the AEC's vast domain. In 1977, when Hirsch left the government, the United States was spending ten times that amount on fusion.

Hirsch was a scientist, but from the beginning he was really an outsider to fusion. He was schooled as a nuclear engineer, not as a plasma physicist. Although he worked in fusion later, his first research job was not with an established plasma physics outfit but with International Telephone and Telegraph's research lab at Fort Wayne, Indiana. He worked there with Philo T. Farnsworth, a great intuitive scientist who was one of the inventors of television. Farnsworth moved on from cathode-ray tubes to fusion energy, tinkering with modest, handmade laboratory devices in the tradition of Lyman Spitzer.

The research did not go far, as Farnsworth's failing health slowed the investigation. But he did impart to Hirsch some essential principles of invention – to attack the central problems and, wherever possible, to work with the real components of an experiment instead of substitutes. In essence, Hirsch learned to keep his eye on the ball.

Hirsch joined the AEC just in time to attend one of fusion's groundbreaking events. In 1968, Hirsch was in the audience at Novosibirsk when Lev Artsimovich announced that his tokamaks had produced record-setting temperatures and conquered the cynical Bohm diffusion theory. When Hirsch returned to his Washington office, he expected the phone to be ringing off the hook with calls from American lab directors begging for money to build their own tokamaks. The phone was remarkably silent.

Hirsch, a medium-sized man with soft features, recalled his decade in the fusion field in an interview years later. While giving a forceful and highly opinionated account of that time, he nevertheless came across as completely reasoned and objective, a talent that no doubt helped in his government role.

What he described was a fusion community that had lost an element of daring over the years. By 1968, Hirsch believed, fusion physicists had retreated from speculative engineering devices back into the safer realm of basic research, to the comfort and pleasure of examining numbers, graphs, and theories. He felt it was a self-defeating prudence, for it would be impossible to take the next step toward a fusion reactor with total understanding.

The British ratification of the tokamak changed that prudence. Within a year, Washington had approved the building of five new tokamaks in the United States. By 1972, there were seventeen tokamaks completed or under construction outside the USSR, including two in France, two at England's Culham lab, and three in Japan. Princeton was building by far the biggest – called the Princeton Large Torus (PLT) – with room for clouds of plasma measuring about a meter across, or forty-five times larger than Lyman Spitzer's original.

As the tokamaks proliferated, Hirsch impatiently pushed on toward the next steps, trying to secure the future of fusion research. In 1970, he launched a campaign in Washington to make fusion research popular outside of scientific circles. He made claims for fusion's potential to congressional committees, to lobbyists, and to the electric power industry.

He urged a massive program for fusion on the order of the Apollo moon mission, a space triumph that captivated Americans in the summer of 1969. With that kind of support, he wagered, the United States could switch on a demonstration fusion reactor by 1995. He won points with the nascent environmental movement, touting fusion as clean energy, and he pitched the fusion reactor as the energy machine that could answer America's anxiety over steepening oil prices and burgeoning electricity needs. Hirsch was completely persuasive. Fusion was cheaper, safer, and less burdened by radioactive waste than the "breeder" fission reactor then favored within the Atomic Energy Commission as the power producer of the future.

For fusion, the timing was exquisite. In the early 1970s, the United States' energy program was flickering and so were the lights. Blackouts and brownouts were household words. President Nixon and officials within his administration, indeed the nation as a whole, were trying to come to grips with a looming energy crisis. A new energy policy was needed to deal with anticipated power shortfalls over the next two decades, a policy that was mindful of the nation's fear of environmental blunders in the hunt for fossil fuels.

Responding, Nixon made nuclear power the centerpiece of his "clean energy" plan. Government and public awareness of fusion's potential grew along with the national debate over the kind of nuclear power to employ. The Nixon administration focused on the proposed "breeder," a variant of the fission reactor that produced more fuel than it burned. The breeder would operate on a mixture of nonfissionable uranium 238 and the naturally radioactive but rare uranium 235. During the fission chain reaction, the stable uranium would be converted into the synthetic element plutonium 239, a highly volatile and toxic radioactive fuel. Controversy swirled around the breeder because of the dangerous volatility of the plutonium, the production of long-lived radioactive byproducts and the fact that plutonium could be diverted to weapons uses.

Fusion stood in marked contrast to the breeder reactor. The fusion reaction itself produces no radioactive waste products (although the interior of the machine does eventually become radioactive). A fusion reactor cannot "melt down" – there is so little fuel in the reacting plasma that it quickly runs out of energy if left alone. Second, the fusion reaction is intrinsically so difficult to sustain that any upset of the process in an accident would instantly end it. In fact, if the reaction gets too hot, the probability of particles hitting and fusing actually goes down. And if the plasma touches the "cool" walls of the reactor vessel, it is quenched instantly.

Fusion's reactor-grade fuel, tritium, a radioactive form of hydrogen is quickly dissipated. Half of any amount of unstable uranium 235 takes an astounding 713 million years to lose its radioactivity. Similarly, the half-life of plutonium 239 is 24,360 years. But the half-life of tritium is only twelve years. In addition, should tritium gas escape, it would go straight up in the atmosphere, like a helium balloon. In the case of a water-borne tritium accident, the half-life for the loss of tritium from the upper layer of soil is a matter of days.[1]

Fusion started receiving favorable notices in the press. In the summer of 1971, the *New York Times* editorialized that fusion stood out as a power source that would "exact minimum environmental costs or none at all."[2] The writer urged Nixon to move constructively to meet the energy dilemma "by substantially increasing the amount now being spent on fusion power research."

Behind Hirsch's boosterism, the scientists knew that fusion's main drawback was rather serious, to put it mildly. No one yet knew how to make a fusion reactor work. Another two decades of research might be necessary. Fusion could not end today's blackouts.

Within the AEC, tension rose over competition for research dollars between fusion and breeders. Fearing that fusion might lose out, scientists at the nation's top fusion labs downplayed the competition. Dick Post, the fusion prophet from Livermore, told a *New York Times* reporter: "We've got two really good horses to ride and we ought to ride them both."[3] He predicted that fusion power would be controlled in the 1980s and that it would be economically "in full swing" by the 1990s. Mel Gottlieb at Princeton was a little more conservative, predicting the completion of a demonstration plant only within the century. It would be a mistake, he warned, to divert funds from nearer-term research.

Bob Hirsch was in the thick of the debate. He was certain that tokamaks were an advanced enough design to bang out big, superhot plasmas. To do so would mean the scientists would have to stop toying with mere hydrogen fuel and start puffing into the magnetic bottle the more exotic forms of hydrogen – deuterium and eventually even radioactive tritium. Deuterium-tritium plasmas – DT, they called it – could produce more abundant fusion collisions, and if properly contained, might be able to reach the reactor range of 100 million degrees centigrade, ten times the level produced by the T-3 in Moscow. It would force the physicists to confront the engineering problems of a genuine reactor, to learn how to handle tritium and how to deliver power to huge electromagnets.

Hirsch also believed that a demonstration of DT burning could be politically useful, providing an impressive demonstration of progress in order to win even stronger government support.

The political atmosphere seemed right. James R. Schlesinger, Nixon's new AEC chairman, was more oriented than his predecessor toward practical goals over general research. It was time to move out of the "pure physics" rut, Hirsch thought, and onto a track of practicality.

"I came into the program with an attitude that I wanted to build a fusion reactor," he said, "but I didn't want to do plasma physics for plasma physics' sake."

In 1972, Hirsch was elevated to director of the fusion branch of the AEC. Later that year, in an assertion of his new power, Hirsch stunned the various fusion lab directors around the country by shutting down two intriguing research machines which nonetheless had not made significant progress toward temperature and confinement goals. He also set concrete milestones for progress, assigning a ten-year timetable at the end of which fusion's "scientific feasibility" would be proved. The goal was called "breakeven," and it meant basically that a fusion reaction would yield as much energy as was needed to create the reaction in the first place.

Hirsch's point had been made. The people who held the pursestrings in Washington were now in charge.

Princeton's Mel Gottlieb retains sharp memories of that time: "Hirsch wanted it understood there was one boss, there was only one God and it was Hirsch. He promptly dissolved the committee of lab people who'd really run the program and substituted his own people so he had a majority in terms of control."

In the summer of 1973, Hirsch called a meeting of the lab directors at a luxury hotel in Key Biscayne, Florida, to discuss a giant tokamak and the idea of burning deuterium and tritium. "Princeton was dead set against it," said Hirsch. "Mel Gottlieb was absolutely shocked; it was almost disdain for even talking about that sort of thing." Princeton, Hirsch felt, viewed itself as a plasma physics lab. "Their attitude was, you had to understand all the plasma physics before taking the next step." Building a machine using deuterium-tritium fuel before other goals were met "never even entered their mind because it was idiotic to even think of doing such a thing in the first place. This was preposterous, unthinkable."

As far as Princeton was concerned, a tritium burn was merely a publicity stunt. Experimentation had not yet begun on Princeton's new $14 million tokamak, the Princeton Large Torus – a machine nearly identical to the Russian T-10 – and already Hirsch was pushing for a bigger machine. "Tritium," said Gottlieb, was "out of the question. If you don't have good enough confinement time, the experiment isn't going to work no matter what you put in. You've got to solve these scientific questions without using tritium."

The inertia at Princeton was thick with scientific arguments and a sense that a bigger machine would merely mean bigger unknowns and possibly even bigger failures. But Hirsch was unmoved. "He just brushed these arguments aside and said, did I have faith or didn't I have faith? Was fusion going to be successful? That was the basic question," Gottlieb recalled. "I said the basic question is, do we have faith that we can go in and get the knowledge required to do it?"

"We didn't speak for some months."

In the Middle East, meanwhile, the year 1973 brought yet another war between the Arab states and Israel. In retaliation for American aid to Israel, the oil-rich Arab nations cut off oil sales to the West. Gasoline lines formed around the country, and Americans were more willing than ever to hear about new energy approaches.

Hirsch continued with his big plans. He encouraged scientists at the Oak Ridge national laboratory in Tennessee to work up a design showing what a tritium-burning reactor might look like. Oak Ridge, with its practical experience in nuclear weapons work, nuclear fission reactors, and the handling of radioactive materials, was not put off by the idea. Hirsch knew the competition would eventually bring Princeton around. While Oak Ridge worked on its design, the AEC and the lab directors around the country continued hashing out the future of fusion.

"Pretty soon," Hirsch said, "Princeton began to realize this thing was getting serious. Gottlieb realized they had to get involved or this thing would pass them by."

Oak Ridge proposed that it build a tokamak equipped with the most advanced and powerful magnets ever contemplated. They would be superconducting liquid magnets. The tokamak itself was huge, approaching the dimensions and magnet capabilities of a reactor. The cost of such a machine would be several hundred million dollars, but it would be capable of skipping the breakeven step altogether and leaping straight to a self-sustaining or "ignited" plasma that could produce excess power.

The very idea that Oak Ridge would land the right to build such a colossus stuck hard in the throats of the Princeton men. Mel Gottlieb was worried for the future of his institution, the Princeton Plasma Physics Laboratory. It was not a multipurpose national laboratory like Oak Ridge, Los Alamos, and Livermore. The government did not have an underlying commitment to keep it open; instead Washington periodically renewed an agreement to have Princeton University operate the fusion facility. Although scientifically Princeton was powerful, "we didn't have a great deal of political clout," Gottlieb explained. "The only way we could get support was just by working harder and doing better." Princeton was in the midst of building the $14 million Princeton Large Torus, but after that the lab's future was uncertain.

Gottlieb wanted the breakeven experiment for Princeton, but he did not want radioactive tritium. Scientifically it was too soon, and it would greatly complicate meaningful experiments. On the other hand, if Washington was set on spending several hundred million dollars to build a giant tokamak, that tokamak had better be at Princeton. At an AEC committee meeting, said Gottlieb, "Furth got up and said, 'Well, if this is what you really want to do, I'll tell you how to do it.'"

Harold Furth may have once misjudged the role of intuition in fusion work, but he was not one to carry that burden long. Indeed, he was the cleverest of men when it came to scientific puzzles. The puzzle was this: how to come up with a competing tokamak design that would advance the temperature, confinement, and density of the fickle plasma without building a gargantuan white elephant. The competing design he proposed incorporated a winning combination of political, economic, and physical factors. Furth's design was much more conservative in size and features than the Oak Ridge proposal, and it drew on what little was known for certain about the tokamak. It held the twin spartan virtues of costing less and risking less. Among the design's trade-offs was a rather weak capability to confine the plasma but superior heating.

Stepping to the blackboard before the delegates from Princeton, rival Oak Ridge, and the other fusion labs, Furth explained his design idea. He called it the "wet wood burner." So long as one held a torch to it, a log of wet wood was capable of burning. The torch, in this case, was a relatively new heating system called "neutral beams" that was being tested on the PLT. After a tritium-fueled plasma was initially warmed by an electric current passing through it, a stream of fast, neutrally charged deuterium particles would be injected into the machine from electrical "beam boxes" outside the tokamak. These fast deuterium particles would fuse directly with the tritium plasma as well as knocking into it and imparting their energy to the plasma cloud. By using neutral beams as an external heating source, the physicists would not need quite as high a level of confinement to reach breakeven. Thus, the machine could be smaller than what Oak Ridge had proposed.

Furth's design embodied the brute-force approach to physics, a stoking of the fusion furnace in pursuit of the high temperatures the Russians were also after. To win the assignment from the AEC, Princeton would agree to eventually burn tritium after all.

Bob Hirsch now had two designs before him – two extremes, really. Oak Ridge had presented what he called the "behemoth." And Princeton had given him the cheaper "gimmick" with a $100 million price tag. When Oak Ridge was asked for an assurance that it could meet the 1976 starting date for construction, considering all the technical leaps it was proposing, the lab director hesitated. Without an absolute guarantee from Oak Ridge, Hirsch turned to the more conservative and cheaper Princeton machine. In the summer of 1974, the Atomic Energy Commission approved the assignment of a new and larger tokamak to Princeton. It would be called the Tokamak Fusion Test Reactor (TFTR), although it would be nothing close to a working fusion reactor.

Oak Ridge was furious.

Once again, the Princeton negativists had won out. In 1969, they had initially built a tokamak only to prove Artsimovich wrong. Now they were about to build the largest tokamak with the biggest plasmas ever seen in the world – but only to satisfy Washington and to ensure the overall health of the Princeton lab, not because they believed that this new machine was in the best interests of fusion research. By 1974, its estimated cost had doubled to $200 million. But if it worked – and if the machine could leapfrog the Russian T-10 and whatever the Europeans might come up with – there would be the prestige of being the first nation to achieve scientific feasibility, or breakeven.

Hirsch was not through. Despite his conviction that tokamaks held the key, he did not wish to eclipse the rest of the nation's fusion research program by betting on tokamaks alone. The mirror machine team at Livermore had been afraid of being shut out by Hirsch's commitment to a DT-burning tokamak. Livermore had been diligently turning out variations on the mirror theme for a decade, making only modest progress in relation to Artsimovich's huge leap. The evidence favoring the mirror approach was scant compared to the tokamak, but, in 1975, Livermore made significant gains in plasma confinement and argued that it was time for an experiment to see if the mirror machine could also reach breakeven.

Hirsch, with a promotion that year that gave him broader control over energy research, was persuaded, and so was his successor as fusion chief, Edwin Kintner. Washington gave Livermore the okay to build a giant mirror machine with a vacuum vessel roughly the size of a Boeing 747 jetliner – more than 200 feet long and 35 feet across – capped by magnets weighing 375 tons each. The estimated cost was just under $100 million.

The fusion community immediately dubbed Hirsch's East Coast–West Coast strategy "the Great Shootout." It was obvious to all that in the next round of reactor building, the government would never be able to afford parallel projects.

Either tokamaks or mirrors would survive. One would be mothballed and the other would drive the United States toward breakeven.

7
Forming the major league

Japan had always maintained a conservative approach to fusion research. But in the 1970s it, too, would venture into the world of Big Science, not with the headlong boldness of a Bob Hirsch but in a style characteristically quiet and effective.

As in most Japanese endeavors, a consensus of the interested parties was considered vital to success. After the fusion disclosures at the 1958 Atoms for Peace conference in Geneva, when other countries rushed to build experimental machines, Japan spent a year in national debate on the best approach. One course was to have the Japanese Atomic Energy Commission immediately build a test machine on the order of Princeton's C-Stellarator and the British Zeta. The second course was to study basic plasma physics at universities, using small machines. The aim would be first to develop an expertise in the science and then to train a generation of young physicists. In the end, Japan chose the university route.

There was a small penalty for this caution. Several of Japan's brightest plasma physicists, who were studying in the United States, had planned to return home with their new expertise but instead found jobs at American labs. Without any prospect of conducting experiments on a large Japanese test reactor, they could not resist the playground of U.S. machinery.

Such was the decision of Shoichi Yoshikawa, a fusion enthusiast who had earned his PhD in nuclear engineering from the Massachusetts Institute of Technology in 1961. His education abroad had been subsidized by the owner of *Yomiuri Shimbun,* a major Japanese newspaper. "My plan was to go back," Yoshikawa recalled in an interview in his office at the Princeton Plasma Physics Lab, "but of course the living standard here is very good and at that time really there was not a major experiment going on in Japan. The very best in the West was Princeton, and so I came here." At Princeton, Yoshikawa was immediately given absorbing work designing a Stellarator alternative. Then in the latter part of the 1960s he played a major role in the design of the Princeton Large Torus. Japan's loss was Princeton's gain.

In the 1970s, after the West's ambitious plans to build giant tokamaks became known – and after the Arab nations slapped on an oil embargo – the Japanese considered taking the same daring leap from their own modest test machines to a colossal tokamak. Once again, the decision emerged from a national consensus carefully nurtured among scientific, political, and industrial leaders.

Tadashi Sekiguchi, a fusion scientist who would later head the University of Tokyo's electrical engineering department, was party to this second debate. "Japanese economics was very upset by the first oil shock of 1973," Sekiguchi recalled. "We realized our energy situation was very fragile. It could easily collapse. Importing raw materials of all kinds, making goods, adding value and exporting is the only way Japan can live."

Fusion presented both a grand opportunity for the Japanese and a large risk because of its expense and the speculative state of the science. As an island nation poor in natural resources and wholly dependent on foreign sources of oil and uranium, Japan was acutely aware of its energy vulnerability. Fusion, the utopian energy from seawater, was seen as a way to guarantee energy independence and security. Also beguiling was the fact that the mysteries of this futuristic science were still waiting to be unraveled. Japan could accomplish much more than merely perfect a proven technology. Japan could lead the way. As a vehicle for promoting national pride and self-confidence, fusion could be important in a spiritual sense, its proponents believed. This yearning to demonstrate Japan's creativity and leadership, to itself and to the world, is often voiced within the Japanese fusion community.

Harold Furth, director of the Princeton University Plasma Physics Laboratory. Courtesy PPPL.

"If the Martians were attacking, if money were no object and the military wanted a working fusion reactor by the year 2000, there is no question we could have it."
—HAROLD FURTH.

Sekiguchi explained it this way:

> We are asking ourselves whether we have a frontier spirit. We are challenging ourselves, asking the question of whether we are qualified. The Meiji restoration was around 1868. There has been only 100 years of civilization of modern science and technology. Almost all the time we have spent bringing things in, catching up. Fusion is a rare example for us to wrestle with unknown problems. It is a quite spiritual issue, a self-image issue.

On a more practical plane, industrial and government leaders saw fusion research as a way to provide engineering jobs and technological spinoffs for Japan's hungry economy. As a postwar pacifist nation, Japan did not have a mammoth weapons manufacturing sector to drive the engine of technological innovation. The engineering and physics requirements of a large-scale fusion machine might fill some of that gap, giving Hitachi, Toshiba, and the other industrial giants a chance to work on major problems of magnetics, vacuum control, metallurgy, cooling systems, and electric power supply.

In 1973, a high-level committee of influential Japanese inside and outside the government convened to decide whether Japan should build a giant tokamak.

"It was like a shadow cabinet," said Kenzo Yamamoto, the government physicist who is credited with shaping the committee and developing Japan's modern nuclear fusion program. It included the vice-chairman of the Japanese Atomic Energy Commission; the directors of several electric utilities; representatives of Toshiba and Hitachi, which had built most of Japan's research machines to date; and even a journalist on the editorial board of *Asahi Shimbun,* the country's largest circulation newspaper. "To get a consensus, newspapers are very useful," said Yamamoto.

A key player was Shoichi Yoshikawa. An insistent and outspoken man, he took a two-year leave from Princeton and returned to Japan for the debate. Experienced in tokamak design after his Princeton Large Torus assignment, Yoshikawa worked with Hitachi to supply the shadow cabinet with the design and engineering requirements for a giant Japanese tokamak. In size, it would be almost exactly the same as the tokamak that Bob Hirsch wanted Princeton to build. Yoshikawa lobbied for national leaders to set aside a substantial construction budget and establish a target no less ambitious than "breakeven."

Meanwhile, he taught applied physics at the University of Tokyo, organized a fusion summer school, and encouraged those studying rarified fusion theory to interact with those working on actual fusion experiments. He wrote Japan's first book for the public explaining fusion and its potential impact on society. He was passionate and emotional about his ideas and, from his point of view, utterly practical.

In late 1974, after more than a year of deliberations, the advisory committee accepted the plan favored by Yoshikawa. The following year, the Japanese AEC elevated fusion to the prestigious status of a "national project," alongside space exploration, ocean development, and fast breeder reactors. This was more than a label; "national projects" were accorded budgets insulated from annual legislative tampering. It took almost three more years to plan the venture and iron out industry contracts, with the government granting final authorization in early 1978. The daringly named "Breakeven Plasma Test Facility" would be built at a government fusion research center north of Tokyo.

The original budget for the machine was about 50 billion yen, as Yoshikawa recalled, more expensive than the projected price for Princeton's new tokamak However, the Japanese would get a larger machine than Princeton's TFTR and one that was guaranteed by the industry contractors to work at the first flick of the switch. The machine components would be pretested in the factories at the full power rating but, even after delivery, they would remain the responsibility of the individual manufacturers until they were working reliably, in concert, in the lab. American scientists were jealous of this arrangement, which they likened to a warranty.

Shoichi Yoshikawa did not come with the package. Although he wanted to remain in Japan to oversee completion of his design, the directorship of the promising new Japanese tokamak project went to another expatriate Japanese scientist, who had also spent the past decade working on American machines. His name was Masaji Yoshikawa (no relation) and his apprenticeship had come at General Atomic in San Diego.

Wounded by the rejection and with a family that preferred life in the United States, Shoichi Yoshikawa left Japan for good. He returned to Princeton to work and to teach, and he watched the building of his machine from afar.

Europe's answer to Lev Artsimovich was even grander than the American and Japanese response.

The Joint European Torus, the colossus in the English countryside, came to be incarnated in a peculiarly European fashion. Bob Hirsch needed only to rally a single government to fusion's cause and keep the American labs from squabbling. The fusion men in Europe were faced with forging a multinational compact, for, in truth, none of the European nations could buy their way into the era of Big Science alone.

In 1969, the year the British laser team confirmed Artsimovich's tokamak claim, Sebastian Pease's brown-brick fusion laboratory at Culham was withering away. A drastically truncated laboratory budget had been imposed by the British government, as it had on all government services. Culham's esoteric work had not been deemed worthy of an exception. The infamous Zeta "pinch" machine was shut down for lack of funds before the scientists had gotten their fill of experimenting on it. But, as the British saw it, with the new results from Moscow, tokamaks were "in." Pinches, Stellarators, and every other shape were "out."

Pease knew that the Americans had converted their C-Stellarator to a tokamak and were building the Princeton Large Torus. He knew that the French had plans for a tokamak of the same size. And he knew that Artsimovich was busy building a bigger tokamak. Opting for larger machines was only a matter of simple calculations, and Pease, the Culham director, wanted Britain in the game. Modern fusion was now well beyond amperes, the standard unit for measuring the strength of an electrical current. Fusion people now talked in millions of amps, in "megamps." If you wanted to build a machine that could come close to producing a self-burning cloud of plasma, you had to think big.

"It seemed to us quite obvious that we needed tokamaks which were carrying megamps," said Pease, "and we knew that the Princeton people were climbing on that bandwagon and that the Russians were going to do it and unless we built a tokamak that could produce the one megamp currents we already had in Zeta, we were all going to be left behind." The problem was money. The hugely expensive steps needed to develop a reactor seemed much more than was justifiable for a nation the size of Great Britain. The country already had a fast breeder fission program to guarantee electricity supplies for some time.

To Pease, some sort of international collaboration seemed essential. Indeed, it might even be appealing to Britain's European allies as a cost-saving step, Pease reckoned. It seemed so obvious. If Britain could conduct a joint experiment with the Russians, it certainly ought to be able to join in one with the other Europeans.

For a time, the idea caught fire. Pease approached his European colleagues with the notion of working together to build a really large tokamak, and he found a ready audience, particularly among the French. The science made sense. The economics of sharing the construction costs were persuasive. Prime Minister Edward Heath immediately warmed to Pease's proposal, the scientist recalled, and soon after his election in 1970 warned his European counterparts that if they did not unite to build this project they all would be "thrashing around once again in the wake of the Americans," according to Pease.

Agreement among the scientists was swift to the concept of a tokamak that could produce plasmas close to reactor quality – hot, neutron-rich plasmas that would help the physicists better understand how a real demonstration reactor might work. A scientific planning committee was organized through the existing bureaucratic mechanism called Euratom, the European atomic energy organization set up soon after the European Economic Community – the Common Market – was created in 1957. Although Great Britain was not yet a Common Market member, it might soon be. And if not, Pease thought, a contract could be worked out between the British government and Euratom to keep the ball rolling.

The idea of a European tokamak must have seemed too good, too promising, too enticing, too big. For no sooner had the idea received the proper nods of approval among scientists than bickering set it. First, the scientists had to settle the question of what to call the project. Originally it was to be named the Joint European Tokamak, but the West Germans objected (and some scientists frowned on including the flagrantly Russian word "tokamak"). Still fans of the Stellarator, the Germans remained skeptical about the tokamak's viability. What if, during the project's construction, the Princeton large tokamak failed or the Russian results were disproven? The Germans wanted a more general and flexible name; the "Joint European Torus" sounded better. (*Torus* was the Greek word for a hollow ring – the primary shape of all circular magnetic systems, tokamaks, pinches, and Stellarators alike.) If the tokamak concept was discredited, the Germans reasoned, the vessel could be quietly converted to a Stellarator without the public fuss of amending the project's name and drawing attention to its change in direction.

So Joint European Torus it became, and the irresistibly high-powered acronym JET was retained.

That question aside, Europe's fusion scientists negotiated over where the design team should set up shop and who should lead it. Culham was eventually chosen as the site, and, adhering to the cardinal rule of balanced national representation, Paul-Henri Rebut of France was chosen as design team director. "The mad Rebut," as he would come to be called, was summoned from Fontenay-aux-Roses, the fusion lab outside of Paris, where he had just finished directing construction of France's first tokamak, to date the most powerful in the world.

Rebut was a creative scientist, one of those rare types schooled in both engineering and physics. He was enamored of building practical things. To move to his new job, Rebut set sail across the English Channel (like William the Conqueror, it was said) on a sloop of his own design and construction. He was a powerful leader in his own milieu and the design team he led made quick work of translating a bold tokamak concept into blueprints.

To the volatile French builder of JET, the conquering of fusion energy was a question of will power. If the scientists will it to be done and fight for it to be done, it can be done, he believed. Moreover he had sometimes felt that it was within his personal power to bring about that conquest.

"Of course you are bound to the law of nature," he lectured an interviewer in thickly accented English not long after JET's dedication. "Nobody can change this law even if they want to change it. This is what physics is about, but at least if you want to obtain something you have to fight for it. You have to fight not only with the government and different people and so on, but you have to fight also with the experiment itself – against the nature – to bring out what you want to show. You have to reveal what is hidden inside the experiment, and this is not without some fight."

They called him "the mad Rebut" because of his intensity and the depth of concentration that made him appear distracted and that sometimes gave him excruciating headaches. Because his thoughts raced ahead of his words, he was difficult to understand in any language. He had glittering blue eyes, a smooth face under a receding hairline, and a flashing grin.

He held strong opinions, which he stated without equivocation. He seemed never to show a trace of self-doubt. He was disdainful of those who were hesitant about building more advanced and expensive research devices.

Rebut's interest in science developed in much the same way as it did for many of the fusion physicists. They all had felt the power and the satisfaction of problem solving from an early age. The more difficult the question, the more powerful they felt when they had solved it. As adults they found themselves drawn irresistibly to what they casually called "interesting questions." Once committed to a challenge, they were absolutely tenacious. To give up was to admit the limits of intellect, and no self-respecting physicist would ever do that.

For Rebut, it began with the electric train set he played with at age ten. He built nearly all of it himself, he said with pride. Not the engine and the carriage, but all the rails, the electrical system, the transformer, the relays and the semiautomatic controls. He had laid multiple tracks, and by the time he was finished with the construction, he could run four or five trains at the same time without collisions. He sternly forbade his younger sister Anne to touch the machinery. It was his creation and domain.

Later, he discovered that as an engineer he rarely needed to rely on mathematical computations to assess the workability of an idea. He could just look at a drawing, he said, visualize the proposed structure and know immediately if it was right or wrong. It was a kind of instinct, almost an "aesthetic" was the word he used. He believed that "the forces must flow in the material like a fluid without a breaking point." The system had to be balanced, integrated, not merely a conglomeration of separate parts that theoretically worked in isolation.

It was this curiosity, brashness, and a sense for "la matiere," as he put it – the substance – that Rebut brought to JET's design team. His confidence, even arrogance about his own abilities, his need for control, and his penchant for being right most of the time made him a formidable leader.

Not everyone shared Rebut's drive and optimism, and he knew it. He complained of a "lack of faith" in some of his colleagues and then said in flat tones, "I suppose that if you asked the people you will get 60 percent, if it is not 80 percent, that will tell you . . . fusion is not possible. Or not impossible, but so far away that they are not able to think about it."

"I am not so interested in the past," he continued, suddenly intense and righteous. "I am not sure even if I am interested in the present. I am interested in the future."

Setting out to design JET in 1973, Rebut envisioned a machine more than ten times the size of the latest French tokamak, a machine capable of up to five megamperes of plasma current – twice what was planned for the American machine. If pushed, this machine theoretically could reach the region of reactorlike, self-sustaining plasmas. But it was not just a bold size that Rebut advocated, it was a bold new shape for the plasma. Tokamak plasmas hemmed in by magnetic fields had been roughly circular in cross-section, a fat tube whirling round and round the ring-shaped vessel. Rebut wanted to alter the magnets' configuration to produce D-shaped plasmas, plasma clouds flattened along the inner edge. It was a new idea that some theorists were pushing, and it was very attractive to engineers; the magnets were much easier to assemble. Rebut was positive it was the right step and was willing to bet several hundred million dollars of Europe's money on it.

In the thrall of Rebut's shimmering intelligence and flinty confidence, the design team executed the plan. Rebut's mandate from the JET scientific committee had been to produce three different designs for JET. Instead, he dared the committee to defy his vision. He produced just two designs, a gargantuan D-shaped plasma tokamak and a slightly smaller version of the same thing. The "smaller" version was so large that three men could stand atop each other's shoulders within its plasma vessel.

The planning committee, particularly its vice-chairman, Britain's Roy Bickerton, was stunned and it rejected the larger tokamak. That left the "smaller" design, and, in May 1975, a comprehensive plan was officially presented to Euratom for approval and site selection. It had taken Rebut's design team just eighteen months to produce the detailed document. The crew expected a quick decision in turn from the Common Market Council of Ministers. The scientists hoped for a decision by the summer. If they got the go-ahead, they thought they could have JET built and running by 1978.

The design team members, particularly the French, were eager to leave England. They had spent almost two years in a miserable, misty land as fusion gypsies, uprooted from their homes, living in rented quarters, their children adjusting to a new language, their wives (some of whom had been employed in their native country) unable to get work. Rebut and his crew were impatient. They wanted to move on to JET's permanent home, wherever that might be, settle their families, and plunge into construction of the machine of their vision. Everything was going beautifully with Rebut at the throttle.

Then the European ministers, guided by political and not scientific currents, took hold of the controls, and JET stopped dead in its tracks.

Like other fusion scientists before him and since, Rebut was to be rendered utterly helpless, becalmed and nearly defeated by a political stalemate. When it came time to decide exactly where JET should be built, a paralysis descended on the Common Market ministers. For two years they were locked in debate during which time Rebut lost half his men and the JET project came within a hair's breadth of expiring altogether.[1]

Euratom, a sort of clearinghouse for European energy grants, was not known for speedy decisions. Action on major questions was taken only after unanimous votes by the Common Market Council of Ministers, whose members changed depending on the issue to be decided. In JET's case, research ministers and foreign ministers had a hand in the decisions. With the prospect of landing a billion dollar fusion research project, each of the ministers found national interest difficult to ignore. No fewer than six sites were proposed for the big fusion machine. The French offered a home for JET at one of their labs, the Germans had two, the British offered Culham, the Belgians had a site, and a failing Euratom-run fission laboratory in Italy was also proposed.

Self-interest took over. In December 1975, the research ministers met for six hours on the JET question but adjourned after asking for a formal recommendation on a site from the European Commission, the EEC's multinational executive body. In January 1976, the commission endorsed Euratom's financially shaky fission lab at Ispra, north of Milan.

When the Council of Research Ministers met again in February, Britain, France, and Germany effectively vetoed Ispra. Stalemated, the council bumped the decision to the Council of Foreign Ministers, who did little more than nod yes to JET but pass the site question back to the research ministers.

At Culham lab, Pease, Rebut, and the design team were at their wits' end. They had written pleas to various government leaders but nothing was effective in moving the JET site decision any closer. Meanwhile, Rebut was losing his precious staff members. They were accepting more promising job offers elsewhere or, sick of being away from home, they were moving back to their native countries. Contract money from Euratom needed to finance the team's work would run out in a matter of months.

Remarkably, no one had argued against building JET. Once the scientists had agreed, the political leaders felt perfectly comfortable with the idea, that is, as long as the pie was properly sliced. British Prime Minister James Callaghan, for example, had always been intrigued by the JET proposal. It had come across his desk when he was Foreign Secretary. Britain, Callaghan recalled in an interview, was having a rocky time during its initial entry into the Common Market, and he saw in JET a way to move Britain off the defensive. By embracing a joint European energy project, a project Britain needed least of all because of its substantial coal reserves and North Sea oil, "it would show that we were ready to make a positive contribution to the future of the community," Callaghan reasoned. "Britain was getting a bad name. I wanted us to look as though we were really trying to make the community work."

When Callaghan learned that Culham was a leading fusion laboratory, he realized the further bonus that the tokamak might be located in England. Hitherto, most European Community projects had meant sending British personnel to the continent. JET could reverse that trend.

Callaghan set about lobbying his fellow European heads of state to endorse JET. He brought the proposal to the attention of West German Chancellor Helmut Schmidt and French President Valery Giscard D'Estaing. "I tried to get both Giscard and Schmidt to see that this was really a twenty-first century proposition and that Europe should not be left behind, and I was the one of the three of them who happened to have stumbled on it and therefore had to alert their interest and support."

Over the course of the two years, while the various ministers remained in a stalemate, Callaghan continued his own JET campaign. "I lobbied everybody. Literally everybody. I made it clear that Germany would have a large share of the work if it came here. I got our ambassadors to visit the Danes, the Irish, the Germans, the French, Luxembourg, Netherlands, everybody I think except Italy." At one exasperating point, he told Schmidt that if the European Community failed to build JET, the United Kingdom would consider independently arranging its own multilateral agreements and build it at Culham anyway; there had been overtures of interest from the Shah of Iran.

Relegated to the political galleries, the scientists saw that the situation had reached a point where most of them did not care which country landed JET, only that it be built, period. Pease retained a preference for having it in the United Kingdom, but Rebut and Bickerton would corner him in the Culham cafeteria, Bickerton recalled, and tell him: "For God's sake, it's better to build it *somewhere* than insist on building it here and have it not built at all."

Out of this almost macabre frustration grew the inevitable mocking solutions. Rebut, only half in jest, suggested building JET on the Queen Elizabeth II ocean liner, docking the project at a different European port each month. The ship's giant propeller could serve the same function, he explained, as the powerful motor generator the tokamak required. The ship could sail full steam ahead and then, with a sudden thrust into reverse, the stored propeller power could be used to pulse the tokamak with electricity.

Bickerton's quick solution was to award the site to the winner of television's annual European song contest.

The Research Ministers met again in October and November of 1976 without any progress. A December meeting was cancelled. European Commissioner Guido Brunner of West Germany declared that the project was "on its deathbed." It was a bitter Christmas present. Inside the fusion community, the coming new year meant only that JET would be built or buried.

The Research Ministers met in March 1977 in Brussels. Now, the European Parliament, another branch of the complex Common Market structure, was demanding a decision on JET's fate. Suddenly a new tactic surfaced. The ministers might agree – unanimously, of course – to allow a majority vote of the nine ministers to decide JET's location.

Italy would not agree until it was satisfied that Ispra would receive a healthy share of Euratom's research budget. France insisted on settling the outstanding question of how JET would be managed (it wanted a measure of autonomy from Euratom). Great Britain would not agree to a majority vote but would agree to what it called a "consensus," an eight to one approval. Thus, if there were at least two votes for Culham, the assignment of JET could not go to another lab. The Germans pushed for a majority vote, feeling confident that their Garching site would win out, but if the United Kingdom insisted on an eight to one consensus to protect Culham, the Germans would protect themselves by insisting on the same. The meeting, which began at 7 p.m., ended at 3 a.m. There was no decision.

By the summer, the competition had been whittled down to a choice between Culham, outside of Oxford, and Garching in Bavaria. Yet the Research Ministers, the Foreign Ministers, and the heads of state were still deadlocked on which side of the channel to select.

The fusion scientists of Europe were stricken by the absurdity of it all. The time had come to start closing down the design team and put the idea of a joint European tokamak to rest. Pease and Rebut, setting in motion funeral arrangements, wrote a letter to the Council of Ministers in which they declared the JET design team work "unmanageable" and authorized the closing of operations by September. They retained a shred of hope, but in good conscience they could not keep their employees' lives in limbo. The scientists' wives had even joined in the plea for a decision, sending their own petition to the council. Their children had to be enrolled in school by September.

But September became October, and still the JET scientists, beyond all reason, hung on in England.

Of course, JET was delivered from a grim demise. The end of the story may have been embellished in the retelling by now, so it should be taken as a bit apocryphal. In any event, a bolt of luck struck in the fall of 1977 amid the terrible circumstances of an international hijacking. It was one of those peculiar, fateful events, like the fatuous Peron claim to a reactor, that seemed to bend fusion's path.

James Callaghan, the British prime minister, had scheduled a one-day meeting with Chancellor Schmidt in Bonn during the third week of October when news flashed of an airliner hijacking. A Lufthansa passenger plane had been commandeered over the Mediterranean by four Arabic-speaking terrorists. Over the next five days Schmidt's government resisted the demands of the terrorists although they killed the pilot and set several deadlines to blow up the plane and its eighty-seven passengers. After a 6,000-mile odyssey, the plane landed in Mogadishu, the capital of Somalia on the Horn of Africa. There, West German commandos stormed the plane, shot the four terrorists and freed all the hostages unharmed. A key weapon in the successful attack was a special kind of "blinding" hand grenade that disoriented the hijackers. The grenades were supplied by the British.

The rescue was seen as a triumph for Schmidt. A wave of relief and euphoria swept West Germany. It was into this atmosphere that Callaghan stepped on October 18, the day after the rescue, for his scheduled meeting. Though relations between West Germany and Great Britain had been cool because of disagreements over Common Market matters, Callaghan was greeted emotionally by Schmidt, who told the English visitor, "Thank you so much for all you have done" in supplying the rescuers' pivotal weapon.

In the generous spirit of the moment, a political opening yawned. The two politicians found themselves in a position to solve some of their nations' differences over Common Market matters without being perceived as meekly granting concessions. Among the handful of agreements to which Schmidt acquiesced was one allowing JET to rise at Culham instead of Garching. In turn, it was implied, a German might be favored for director of the project.

A quick meeting of the Council of Ministers was called for the following week. The depleted design team at Culham dared not allow itself to hope too much. The Council of Ministers had met more than ten times in the past two years and failed to select a home for JET. On October 25 the council gathered in Luxembourg. At Culham, Pease, Rebut, and the other scientists waited nervously by the telephone. The call came just after noon. Acquiescing to a majority vote, Europe's Council of Research Ministers had selected Culham as the site for the Joint European Torus.

A champagne celebration erupted among the design team workers. It had taken five years to settle, but Rebut had not lost his faith. He was the sort of person who believed right would always win out in the end. News of the decision traveled fast around the globe as well. The Italian Enzo Bertolini, a JET design team member with a speciality in power supplies, recalled that he was in the United States at a symposium on fusion engineering in Knoxville, Tennessee. Just before giving his talk on the phantom project that was JET, Bertolini was called aside to take an overseas telephone call from Luxembourg. He hung up and then took his place at the lectern before the largely American audience. He was so overcome with emotion he could barely speak. "JET has been approved," he announced, and the audience of 1,000 scientists burst into loud applause.

When it came time for the Council of Ministers to choose a leader for the Joint European Torus, they passed over the brilliant but blunt-spoken Rebut and cast their nets for a bigger fish, a West German, Hans-Otto Wuster, a deputy director of CERN, the European particle accelerator project that straddled the French-Swiss border. Wuster had run a famous international project with a terrific public image that produced work of Nobel Prize quality.

Rebut, who had coveted the directorship for himself, was offered the position of Wuster's deputy, in charge of constructing the JET project. The Frenchman was crushed, insulted, furious. He threatened to quit. But he took the job and, together with Wuster's administrative talents, brought JET in on time and on budget seven years later.

In April 1984, the JET dedication ceremony attracted a full house of dignitaries, most notably Queen Elizabeth and French President François Mitterrand, who was making his first trip to England as president. "All of us here today," said the Queen, "may have a tale to tell our grandchildren when we say we attended the unveiling of JET and witnessed the beginning of the new technological advance. In an energy-hungry age JET may be a step along the road to unlimited electric power."[2] The two leaders noted that it was the eightieth anniversary of the signing of the Entente Cordiale between Britain and France. Mitterrand said that it was on the solid foundation of such agreements that the European community could be built.

Along the wall that leads to the JET control room, past the photographs of Queen Elizabeth and President Mitterrand, one comes upon an inscribed stone that reads: "This foundation stone was laid on 18 May 1979 by Dr. Guido Brunner Member of the Commission of the European Communities." Scientists familiar with JET's turbulent birth often say, as they chuckle irreverently, that there should be a second commemorative stone: "With Thanks to the Four Terrorist Hijackers."

8
The political plasma

The new recruits to the fusion cause arrived at the Princeton Plasma Physics Laboratory eager to get their hands on a big, important machine. They sported fresh PhDs from Harvard, Columbia University, MIT, the University of Wisconsin, Princeton, and other robust plasma physics graduate programs that had benefitted from the enthusiastic Bob Hirsch era.

The recruits had signed up for plasma duty rather than other pursuits in the world of physics because they saw an opportunity for ground-breaking work and improved job security. Fusion was now a field to which Washington had committed long-term resources. Princeton also welcomed rotating contingents of promising young Europeans looking for experience on a tokamak of significant size – and one with neutral beams – before they tackled the operation of the nascent JET.

The new plasma physicists were, by and large, a more practical and cynical generation than the original fusion fraternity. They were sensitive to environmental issues, and while they acknowledged fusion's utopian promise, it was mostly a promise to them and not a life's ambition. They were more wrapped up in the physics at hand, concerned with maintaining a set of interesting problems to occupy themselves. They hoped for fusion's eventual commercial success, but they had more patience, less urgency, and seemed to find satisfaction in interim goals. Still, they shared with their mentors an unflagging determination to conquer the recalcitrant plasma.

Among this generation of physicists was Rob Goldston. When, in 1975, experiments on the intermediate-sized Princeton Large Torus began, Goldston was only twenty-five years old, just completing his PhD at Princeton, where his professors included Harold Furth. Goldston had written his thesis under the supervision of Harold Eubank, who was Princeton's expert on neutral beams.

Goldston already had the look of an esoteric physicist. He wore a black Cossack mustache and bushy black hair that grew out and up so that the circumference of his head increased as the weeks went by. The soft skin of his face and arms was a floury white as though he had never felt the sun but instead spent all his days in a room with a computer, which was largely the case. Still, his intellectual bravado expressed something else about him. He was supremely confident that the human mind could lick fusion.

Like many of his colleagues, he tended to regard the plasma in anthropomorphic terms. It was, Goldston sometimes thought, "a very mysterious and tricky being." But how had he become a student of plasma in the first place?

On Labor Day 1970, Goldston and a friend were standing on the summit of Mount Mansfield in the Stowe region of Vermont. Earlier that day, they had decided to climb the mountain while under the influence of a mind-altering drug.

As Goldston tells the story, he became fixated on the notion of gravity, running up and down the mountainside exulting in the feel of gravity's pull. He shook the trees and watched the leaves fall. Looking out over Stowe Valley he saw smoke rising from a smokestack and remembered that people merely borrow energy as it passes into other forms. We do not consume energy so much as convert it, from the solid form latent in coal, oil, and firewood to forms of heat and light and byproducts that contain further energy locked in their chemical bonds.

In his revery, it struck Goldston that the trick was to get energy, as he phrased it, "without bullshit" – without the smoke that ruined the landscape. To get clean energy in some direct and beautiful way straight from the atom itself, that was the challenge. An article he had read had described what seemed to him to be a promising method. It was called nuclear fusion.

What drove Goldston and many fusion scientists forward in fusion

Interior view of Princeton's TFTR.
Courtesy PPPL.

was the thought of delivering a gift of virtually perpetual, clean energy to the world in the face of mounting scarcity. When he came down from the mountain, Goldston set out to learn the science of plasma physics.

Goldston and his peers had excellent timing. The fusion scene they entered offered tremendous opportunities for discovery plus the spur of close international competition. It had been more than a decade since Artsimovich had introduced the world to the tokamak, yet the scientists were still trying to understand why the Russian doughnut-shaped device worked so much better than other fusion schemes. And because of the rapid buildup of new and bigger machines there was much experimental work that had to be done quickly before the next machine was ready for investigation. For the community, the proliferation of expensive tokamaks meant that developments in one country would have a direct bearing on work elsewhere. Lab directors kept close tabs on decisions in rival labs, hesitating to modify their machines until the results from other countries' tinkerings were in. Collaboration increased because of the similarity of interests, but labs jealously guarded information with an eye toward being the first to announce breakthroughs that could be duplicated elsewhere. A grand competition was taking shape.

The new generation landed at Princeton in the midst of a dramatic expansion of workload. The Princeton Large Torus was just gearing up in the mid-1970s and its assignment was weighty – to demonstrate whether high-powered neutral beams could indeed significantly boost plasma temperature without spoiling the confinement of the particles. Harold Furth's design for the next step of a giant tokamak, the "wet wood burner," counted on the success of neutral beams. JET and the Japanese tokamak were also depending on neutral beams. A failure on PLT would presage doom on the giants to come.

PLT struck the newcomers as an extremely large machine, Goldston remembered. It was, in fact, the largest tokamak yet constructed in the world, and, at $15 million, it was the United States' premier fusion research device. With all its components it weighed 150 tons and the volume of its vacuum vessel was 221 cubic feet. Its most formidable feature was the bank of newly invented neutral beam boxes, supposedly capable of pouring three million watts of power into the plasma in the form of speeding, neutrally charged ions.

There was plenty of work at Princeton. PLT was to have more than thirty different diagnostic devices to check plasma characteristics. Each device had to be designed, built, installed, monitored, and its readings analyzed. On top of that, the new neutral beams would require special attention. Experiments and maintenance on the machine would frequently run from 8 a.m. to midnight, Monday through Saturday, creating enormous and constant manpower needs.

As in the 1950s, there was an expectation of imminent discovery. But it never stopped being frustrating work.

"It always ran better at night," Fred Dylla, a vacuum specialist, remarked ruefully. "It was a night person. It had the day to work out problems and then it settled into a rhythm."

One theoretical stumbling block concerned the relationship between plasma temperature and its confinement: It just was not fully understood. The conventional and pessimistic line of thought was that as one raised the temperature on a tokamak, plasma confinement would probably deteriorate and reach a point where further temperature boosts would not improve fusion power production. The energy would be lost to the cooler vacuum vessel walls by speeding, escaping particles. It was also speculated that the auxiliary heating of neutral beams might exacerbate the situation. Discovering the relationship of confinement time to high temperature seemed crucial.

Although PLT experiments began in 1975, it took several years to attach the neutral beams. The scientists described the construction pressure that Washington exerted on Princeton as worse than anything they had experienced. The new Cabinet-level Department of Energy had an office in the Princeton lab itself, and there were frequent visits from Washington bureaucrats demanding that Princeton have the four beam boxes working by the end of 1977 – the construction starting date for Princeton's next, giant tokamak.

It was a schedule that Harold Eubank found impossible to meet. He remembered one meeting with the Energy Department representatives that he and Mel Gottlieb attended in which the 1977 deadline was pressed upon them: "I said 'I can't do it,' and they said 'You have to do it.' I'm in front of the room defending this thing when Gottlieb put his arm on my shoulder. You know you've lost when he does that. It's like the kiss of death. 'Can't you get three on?' he asked. I said sure. We couldn't stand there arguing all day, but we got the last two on in '78 – like I said."

While PLT waited for its neutral beams, machines at other labs had been adding to the fusion numbers race. At the Oak Ridge national lab, which had become expert in neutral beams and had actually built the Princeton beams under contract, the small Ormak tokamak had used a neutral beam torch to create ion temperatures of 20 million degrees centigrade, three times what Artsimovich had produced in T-3. Also, the little Alcator tokamak at MIT, using extremely powerful and compact magnets, was showing improved confinement and density numbers at low temperatures in a small plasma.

By the summer of 1978, all four of PLT's neutral beams were in place, and the acid test began for the new heating scheme on a big tokamak. Goldston, who had become expert in analyzing the properties of beam-heated plasmas, was given responsibility for coordinating many of the experiments. As it turned out, the plasma did not react quietly to the neutral beam torch, but its shimmies and shakes were minor enough to be controlled. As the plasma temperature shot up, the confinement time remained good enough to make neutral beam heating worth it. Over the course of just a few weeks in July and August of 1978, the PLT team coaxed more than 2 million watts of power out of the neutral beams and brought the plasma temperature higher and higher until one of the most reliable diagnostic devices put the temperature at an astonishing 5.5 thousand electron volts or about 65 million degrees centigrade.

PLT's temperature record was a huge leap forward in more than one respect. The achievement turned pessimism into optimism for the next giant neutral-beam heated tokamak that Princeton was to build. But more important, it lent new credibility to the idea of creating a net surplus of power from a fusion reaction and the interim goal called "breakeven." Word of Princeton's record promised to polish fusion's reputation at a time when alternatives to nuclear energy were gaining attention. Mel Gottlieb called his Department of Energy overseers to arrange a news conference.

But Gottlieb was startled to learn that Washington wanted no part of the news. Not at that moment, anyway. It was August, and the timing of the Princeton breakthrough seemed suspicious to the Department of Energy, which noted that it came just as Congress was dealing with budgeting decisions. The Energy Department requests for appropriations were based on President Jimmy Carter's new conservation strategy to the energy crisis. The policy emphasis was to be on nonnuclear, renewable and synthetic resources, solar energy, and the development of the nation's coal reserves. James Schlesinger, Carter's new energy secretary, did not want Princeton announcing great gains in nuclear fusion. It would take the edge off the crisis-driven solutions the administration was offering. People might think the nation's energy problems had been solved by the utopian promise of fusion.

Princeton was forbidden to broadcast its achievement.

But word leaked out. Before government officials vetoed the idea, the Fusion Energy Foundation, an independent advocacy group, had learned from Gottlieb that there would be a news conference. Once informed of the cancellation, the group told a Knight-Ridder newspaper reporter of this unusual turnabout. The reporter was able to confirm from other scientists at the Princeton lab that some sort of breakthrough had occurred – one that the Department of Energy apparently did not want publicized. On a slow news weekend, the Princeton fusion achievement was reported nationwide on the Knight-Ridder news wire and picked up by the television networks and other newspapers. Schlesinger and the fusion division heads were furious. William Bowen, then president of Princeton University, who knew Schlesinger personally, was called in by the lab to help calm down the secretary. Eventually, since word was out anyway, the Energy Department allowed a low-key news conference.

As in the secrecy era and at the Geneva Atoms for Peace conference, fusion continued to be an issue fraught with political overtones. To simply pursue the science and think that it could speak for itself was only inviting trouble. Princeton managed to paper over the bad feelings in Washington this time, but it learned a lesson in political prudence that would affect future announcements.

As important as the Washington reaction was to the political fortunes of the Princeton lab, it had little to do with Princeton's scientific reputation. That was an issue for international colleagues. So with PLT results in hand, Goldston and his mentors set out for the August 1978 Plasma Olympics in Innsbruck, Austria, prepared to claim the world temperature record away from the joint holders, the Oak Ridge lab and the French national lab. PLT had delivered plasmas three times hotter than the previous high, but the very magnitude of the increase led to international skepticism. At Innsbruck, it was Goldston's defense of the Princeton paper on PLT's temperature record that made the young physicist's reputation.

"Fry Rob on one side – flip him and fry him on the other side," was how Goldston remembered the questioning from the multinational cast of knife-throwers that attended the special night interrogation organized by Roy Bickerton, the scientific leader of Culham lab in England. That same morning, as a courtesy, the sardonic Bickerton had handed Goldston a list of twenty questions he had gathered, all probing the validity of the Princeton PLT experiments and the 5.5 thousand electron volt ion temperatures. "If we came up with three or four . . . okay," said Goldston, "but five and a half? Everyone thought it couldn't be right."

The Princeton mentors rallied around Goldston as fight managers pump up a contender. Through the day Furth and others coached and rehearsed Goldston as he shadow-punched his way through the list of questions. When new questions from different and unexpected angles were thrown out, Goldston defended and counter attacked. After a time, Goldston's mentors pronounced him ready for the evening bout.

There were several hundred people in the audience, but Goldston said he mostly felt the penetrating mind of the eminent Bickerton. Moving through the list of questions, Goldston was able to stay on his feet as the lances fell. They questioned diagnostic methods, calibrations. . . hours went by. Goldston was answering the seventeenth question strongly, but he was worried about the eighteenth. He knew he did not really have a good explanation. As Goldston was about to offer a soft return, Bickerton interrupted. "Look, we're exhausted," the Englishman said. "This has been going on too long. Thank you."

Goldston and Princeton had made it. The community accepted their work. The new temperature record of 65 million degrees would stand. Within the year, PLT would soar to 82 million degrees. Princeton and the international fusion community were ready now for the next, long-awaited step. It was time to stoke up the giant tokamaks and reach for breakeven.

There was at least one spiritual moment on the day ground was broken at Princeton in 1977 for the next-generation of tokamak that might lead to fusion's promised land. Among the ranking dignitaries was Dr. Yasuhiko Iso, a physicist directing Japan's plans for a parallel energy project. Iso's gift to his Princeton friends was a large papier-mache sculpture, a Daruma doll.

The "doll" was actually a menacing-looking thing, a sort of Kabuki version of Humpty-Dumpty, all head and body, no legs or arms, painted in garish red and black. Under dramatic eyebrows were white blanks where the eyes should have been. The Princeton scientists, instructed by Iso, painted in a single black, beady eye. According to Japanese tradition, the second eye could only be painted when the enterprise begun that day was completed.

The symbol from the East held meaning for Princeton. Daruma, the founder of Zen Buddhism and inspiration for the charmed doll, was said to have spent nine years immobile, meditating in a cave. After some time, his unused legs and arms shriveled and disappeared. But he kept to his task and eventually was rewarded with divine insight. His life became a lesson in steadfastness of purpose, a lesson the fusion community had taken to heart.

For eight years the baleful beady eye of Iso's Daruma doll looked upon the Princeton workers from a case in the laboratory's carpeted lobby as they built the $314 million Tokamak Fusion Test Reactor (TFTR).

Building TFTR proved to be the most intense activity in the Princeton lab's thirty years of existence, although in the town, dominated as it was by the stone spires and ivied walls of Princeton University, the project remained a footnote. More than 1,200 physicists, engineers, and support staff worked at the lab. The collection of tan, smooth-paneled buildings was all but invisible, hidden behind a thick swatch of woods and backed by the grey lattices of electric works and high-voltage equipment fenced with barbed wire.

Princeton and the U.S. Department of Energy had agreed on a deadline of Christmas 1982 for TFTR to begin operations and create its "first plasma." If all went well, the American colossus would be the first of four giant tokamaks to attempt a plasma. JET and the Japanese tokamak were still under construction. The Russian entry was just in the planning stages – and, in fact, would never find the financing to be built.

The men who worked frantically to piece together TFTR remembered those last days and weeks before Christmas as a time of exhilaration and exhaustion.

Don Grove, the chief architect of the big test reactor and the project director, had been working something close to a triple shift to prepare for the machine's maiden voyage. His hours were strictly against the lab rules. He would show up just after midnight and work until 6 a.m. Then he would go home at dawn for a couple hours' rest and report back at 8 a.m., working a full day until dinnertime. He would eat, go to bed, and rise again at midnight.

Some had voiced grave doubts about reaching "first plasma" on time, among them Harold Furth, who had ascended to lab director after Mel Gottlieb's retirement. Furth had issued a decree that no one was to work more than an eighty-four-hour week. But people willing to work eighty-four hours were not stopped at eighty-five. Young technicians, encouraged in part by Grove's example, simply ignored Furth's warning.

Don Grove was not a terribly patient man, but this project had forced patience upon him. More that eight years of toil had gone into TFTR since it was first assigned to Princeton. "As I came down to that wire," he remembered, "I certainly was willing to listen to anybody because I needed all the advice I could get, but I was also beginning to make it my show. I was going to do this."

The men who fashioned TFTR were greeted each morning by the Daruma doll and the stinging smell of the custodians' disinfectant that pervaded the lab offices. To reach the machine, they moved down a concrete stairwell and then used an electrically coded passcard to unlock heavy doors that opened to a long tunnel. The underground tunnel wove the length of two football fields out to what the physicists called the machine test cell, though the "cell" was as big as an airplane hangar. If they were lucky they would find a bicycle at the tunnel entrance and could pedal down the dank, echoing corridor, taking care to press the little metal bell on the handlebars as they came to blind corners.

The tunnel yielded to another concrete building, where the men would climb two flights of narrow metal stairs into what looked like a warehouse. Through a pair of giant, thick doors, they would enter the great hall, the test cell, to find the flagship of America's fusion program waiting to be switched on.

It looked like a plumber's nightmare.

Despite its size – 30 feet high and weighing more than 700 tons –

TFTR seemed like a huge contraption, hastily fashioned in an oversized garage. A conglomeration of pipes and wires were stuck onto a central steel chamber with platforms and outcroppings piled high on top of each other until the machine had swelled to the size of a church. A little American flag poked out from the topmost platform. Men crawled on the outer hull making last-minute adjustments and testing equipment.

The test reactor was the endpoint for what seemed like miles of electric power equipment, water-cooling systems, and computer cabling that started on entire floors above and below the machine. Like turning over the engine of a new car, the "first plasma" would tell the Princeton physicists whether the machine really worked. It was this deadline that had turned Don Grove's life upside down and transformed the laboratory into a sweat shop.

The deadline had been pushed back informally several times, but now the Department of Energy had been sent Princeton's promise in writing, and Congress had been notified that the machine's debut would be in December. Moreover, when 1,200 fellow physicists gathered for an annual convention in Washington the previous April, Furth had announced that the new machine, already a year late, would be completed by Christmas. There was no retreating.

Don Grove had joined the Princeton lab in the Spitzer era, overseeing the birth, life, and death of the C-Stellarator. He had directed its transformation into a tokamak and had supervised the building of the record-setting Princeton Large Torus. The Tokamak Fusion Test Reactor would be his swan song.

Grove had turned sixty-five years old the year of TFTR's "first plasma" deadline, but he only looked like an old man when he was tired. He had a strong, aquiline face with the silver hair thick and combed back, his pale blue eyes always alert, the neck pivoting, birdlike, toward a speaker, for Grove was hard of hearing but could read lips expertly.

The loss of hearing was not due to advancing years but a childhood illness. He had spent his school years nearly deaf, in fact, and had to teach himself physics because his professors kept turning their backs to write on the blackboard, leaving the lip-reading boy out of the academic conversation. It was not until his mid-fifties that he got a high-powered hearing aid and regained some measure of that sense. But he had made do all that time and seemed to have turned the handicap into a mere inconvenience.

Grove believed the lab could produce TFTR's first plasma by Christmas, but he was virtually alone in his conviction. As December 23 arrived, Grove's team had not even tested the power systems that would charge the magnets and heat the hydrogen gas into a plasma. With the pragmatic acquiescence of Furth, Grove, had compromised on this and every other system feeding the machine. The electrical system designed to deliver a huge burst of power for ten seconds was ready to give Grove only a fraction of that power for less than a second.

The multimillion-dollar computerized, automated control system would not be ready in time. So the engineers had rolled some consoles down the long tunnel and set up a temporary control room, just outside the giant double doors of the machine test cell. It was nothing more than computer cases arranged in a square, surrounded by a snow fence. Instead of the intended computer graphs and automatically stored readouts, antique oscilloscopes would suffice. Only the Russians still used such instruments.

People out of the central action remembered hanging on the snow fence, looking in at Grove and his trio of deputies, Dale Meade, Jim Sinnis, and Rick Hawryluk, who were making and enforcing most decisions. Every move had to be outlined beforehand on paper and checked against government guidelines. The secretaries were working hard and fast to keep the official paperwork ahead of the technicians' wrenches. From time to time Grove would sprint down the tunnel to the main building with handwritten paperwork for his secretary, Kay Collins. "Better hurry up!" he would tell Kay with a crafty look. "They're doing it already."

The crews on the new machine had already volunteered to give up the traditional office Christmas celebrations. And Grove had issued a memo warning that everyone involved in first plasma "must strictly abstain from parties" elsewhere in the lab. He did not want a drunken worker to trip up his show.

Grove had even posted a guard at the entrance to the tunnel that led to the test cell. He had instructed the sentry not to acknowledge that he knew anyone on sight but to bend over the worker's identification badge, meanwhile sniffing for any alcohol on the breath. Anyone who had been drinking was to be turned away, no exceptions.

The rush for first plasma had become a test of physical stamina, Graeme Tait recalled. Tait was working more than eighty-four hours a week, but Grove did not try to slow down the skinny thirty-four-year-old Australian. Tait was a shrewd physicist who could understand engineering and power supplies, computers and cooling systems in addition to the complex physics his job description required. He had the big picture.

For nearly two months Tait had spent virtually all his waking hours at the laboratory, and he did not think he looked any the worse for wear. He was a bony man to begin with, his hipbones evident under his slacks, his shoulders poking at the corners of his shirts. Tait was one of many "foreigners" at the Princeton project, the fusion gypsies, hopping from lab to lab, picking up new languages, discarding accents until they could not really say where they came from or with which country they identified. They gave their loyalty to fusion and to whichever machine they were working on at the time.

In spite of himself and his severe misgivings about this big rush to a self-appointed Christmas deadline, Tait remembered being quite enthusiastic about Princeton's latest fusion machine, and he found he could not deny the infectiousness of Grove's determination. Tait was supposedly alternating shifts with Charlie Neumeyer and Dick Cassel, two engineers, but out of anxiety and curiosity the complete triumverate always seemed to be there. Their job was to get Grove the various forms of power he needed to charge up the hydrogen gas and turn it into a plasma.

They had fallen so far behind schedule on the power supplies because the low-bid contractor that built the systems had gone bankrupt before finishing the job. The lab's own staff of engineers had had to modify and complete the power systems.

With each passing week, however, Tait said he grew more cynical. It had become obvious to him, he said, that Grove was not interested in getting the power supplies into the kind of reliable shape needed for the years of experimentation ahead. Corners were being cut in all phases of the operation to focus on the immediate goal. "I was pretty disgusted by that point," Tait recalled, "and it was probably just partly being tired, but also I'd seen so much of this compromising of everything for the one narrow goal of making the sort of flash-in-the-pan first plasma, the political plasma, and it was something that deep down I never agreed with."

Others in the cadre of all-night workers never stopped to question the methods. Grove found it impossible to stop Bubba Vinson, a strapping young technician who sported a navy blue billed cap with gold oak leaf clusters, like some kind of military commander. Vinson was attached to another part of the laboratory, the shop that had welded the machine together, but Grove had approved giving him a chance to be on loan to the "first plasma" crew. He was assigned to install a gas analyzer that would tell the physicists what kind of unwanted gases were in the machine contaminating the hydrogen fuel.

Vinson saw the assignment as his ticket out of the vacuum shop and into bigger things at the lab. Working on the new machine was "the ultimate," "the front lines," he recalled. Back in his hometown of Memphis, Tennessee, he had been a carpenter, a forklift driver, a silo repairman, a gas station attendant, and a steel mill worker. He was through job hopping now. After seven years Princeton had grown to be more than a job, and he was determined to make it here. Like many of the men on the support crews, he said he felt he was working on something important. It was not only scientists who were wrapped up in the fusion crusade. At other jobs, Vinson's labors benefitted a profit-making company. This job, he felt, had a patriotic feel to it. The "highlights," he said, were greater "mainly because they mean more to the American economy and the American people than those things I built before."

That is probably why Vinson worked thirty-nine hours straight on a balky piece of vital engineering before Jim Sinnis, one of Grove's deputies, told him to get out of the lab and get some sleep. Vinson put him off for a few hours but then agreed on one condition, that Sinnis call him just before first plasma was ignited so he could witness the big event. He extracted a similar promise from some of his buddies on the first plasma crew and headed home.

Graeme Tait and company thought their job was about through, too, until 5 p.m. on December 24, when a portable circuit tester indicated that the insulation on a key electrical system was leaking power somewhere. It took several hours to track down the leak, but before it was fixed Grove felt tempted to try the high-power test anyway, a test required by the government before TFTR could be turned on.

Grove and Tait remembered the moment well. The project director asked Tait what he thought about a quick high-power test. The gaunt Australian exploded. They would risk a full scale ground fault, damage to the machine, hours, if not days, more of work.

As Grove recalled, Tait snapped to him: "You're a fool if you do that," and then stalked away to his computer console. Grove decided to fix the power leak.

Furth had watched the whole enterprise with growing anxiety. It was nearing midnight, the technicians were tired, and Grove, he remembered, looked terrible. Furth had wanted an on-time first plasma as much as anyone. He had even iced a case of champagne in the lab cafeteria. He recalled feeling the weight of his overall responsibilities as lab director.

"There was a real hazard, in my mind, that something would go wrong because people were tired," he said. "They might make a mistake, and somebody gets hurt." Furth confronted Grove and told him to give up – they would try next week.

But Grove was hardly a quitter. Reacting entirely in character, he recalled, he told Furth that "if he didn't get out of there I was going to get two big technicians to take him out and tie him to a tree." Furth agreed to give Grove and the technicians the time they needed to make one last try at fixing the power break. But he told Grove that when the white-faced clock over the computer cases struck 2 a.m., on Christmas Day, that would be it.

With one hour to go before Furth's new deadline, the technicians were still sweating over the leak. A crowd had gathered on the snow fence around the makeshift control room. The physicists, the administrators, and the Department of Energy chaperones on hand for the hoped-for first plasma stood by helplessly. Grove and his chief technician, Marty Perron, were staring at the operation when a quick solution for the leaky insulation hit them.

By that time it was already five minutes to 2 a.m. Grove said he looked at his deputy, Dale Meade. Meade looked back at Grove. Then Meade casually walked over to the white-faced wall clock and pulled out the plug.

With some borrowed and jerry-rigged equipment, the insulation problem was wrapped up and the power systems were tested, one by one. Grove pulled from his pocket the government form, which he had already signed, signifying that all the required tests had been completed. A videotape cameraman on hand for the historic night captured Grove slapping down the paper in front of a Department of Energy official, who signed and said, "It's yours," signifying that Princeton now had complete control over the machine.

"Okay let's go!" Grove yelled. "Put the gas in!"

One of the senior physicists picked up a microphone and began an echoing countdown. "Ten, nine, eight . . ."

Eyes swung over to the plasma television, the camera focused on the porthole of the vessel, and to an oscilloscope that would show a green trace if the magnetic current pulsed strongly enough. Marty Perron stared at another television screen, his hand on a palm-sized red emergency stop button, poised to jam it down if he spotted any leaky electric arcing off the patched-up power system.

Out at the power supply room Graeme Tait listened to a headset for the results. Overcome with cynicism, he said, he found himself hoping the test would fail.

"Seven, six, five . . ."

The motor generator sets were whining with power like great jet engines. Grove remembered that he could hardly contain himself, muttering, "Go, baby, go."

"Four, three, two, one."

There was silence and then, suddenly, a flash of light on the television screen. That was it. A yell went up from the crowd.

Grove jumped and punched the air three times with his fist, a gesture broadcast later on national network news. The technicians clapped and whistled. The largest fusion machine in the world had, for an instant, worked. Now there was hope for giant tokamaks everywhere.

On the computer cabinet the white-faced clock was frozen at five minutes to 2 a.m. December 24, 1982. They had made it. The first plasma of the giant Tokamak Fusion Test Reactor had been a poor plasma, of course, so cool that no fusion reactions could come out of it. It was a very low current, just 51 kiloamperes, when ultimately the lab would be making plasmas in the megampere range – fifty times stronger. And the gaseous cloud had lasted for just 50 thousandths of a second, a mere blip on the oscilloscope. The physicists knew that they would need plasma lasting several full seconds before they were through.

Grove's team turned off the power supplies, shut down their computers, and reconvened in the laboratory's big carpeted lobby. Furth broke out the champagne and, like a proud father, passed out silver-wrapped cigars marked: "It's a plasma!"

At about 4:30 in the morning – it was Christmas Day – Grove and Meade made gleeful wakeup calls to Energy Department officials in Washington and trumpeted the news.

An exhausted Bubba Vinson was sound asleep when the incipient new sun had flashed in the machine. No one had remembered to call him. He recalled waking in the morning, looking at his bedside clock and bolting upright. He reached for the phone and dialed the lab. Yes, he was told, it was over. They had made first plasma at 3:06 that morning.

Grove finally drove home but said he could not sleep. At about 7 a.m., he received his first congratulatory phone call, from Japan, where news had already reached his counterparts building the giant Japanese tokamak. The fusion world was very small indeed.

A few months later, the new Princeton tokamak was formally dedicated. Taking in hand a brush dipped in black paint, a familiar Japanese visitor, Dr. Yasuhiko Iso, approached the menacing Daruma doll and painted in the second unblinking eye.

9
The modern fusion lab

The days of the solitary scientist working quietly in a book-lined office, tinkering with a machine when he or she felt the inspiration of a new idea, those days were gone forever from the Princeton lab. The giant tokamak had done that. What was once a quiet art was now frenetic Big Science, and it had transformed Princeton into a modern plasma physics research establishment – a generation beyond Spitzer's intimate rabbit hutch.

Squads of people were now involved in every aspect of TFTR, from the overseeing of the power supplies to the programming of computer controls to the decisions on which experiments might prove most useful. There was an endless schedule of meetings to coordinate the work among engineers, programmers, physicists, and budget watchers. Furth went so far as to bring in a deputy director from the NASA space shuttle program strictly to oversee budget, construction, and personnel matters as opposed to the scientific end of the operation. The machine had initially cost $314 million, but additional heating and power supplies, modifications, and operating expenses brought the real cost of the machine up to about $1 billion – close to the overall cost of building and running JET.

Although the lab would run round-the-clock maintenance and experimental shifts on the machine, there was never enough time to perform all the experiments that were proposed nor enough money to cover desired modifications. Oldtimers reminisced about the Spitzer days, when everyone at the lab could meet around a single table and

Interior view of the Joint European Torus (JET) in Oxfordshire, England. Courtesy JET Joint Undertaking.

decisions could be taken quickly by the people who actually built the machine. Back then it was not unusual for physicists enjoying an evening beer to hatch an experiment and rush back to the lab, switch on the machine, and perform the work that very night.

With TFTR, however, the proprietary sense got lost in the crowd. The machine did not belong to its makers and handlers but rather to the bureaucracy. When the old C-Stellarator was dismantled in favor of a tokamak, some workers were said to have grown nostalgic over the loss. It would be hard to imagine any tears being shed for TFTR when its time was ended.

Rob Goldston's job in the Big Science era included leading group sessions among the physicists working on the new fusion machine. He sat at the head of the long conference table ringed by several rows of men in shirtsleeves. (Princeton, like many American research labs, suffered a dearth of female physicists. Only one worked regularly on TFTR problems.) The young crew, baptized on PLT, had feared that work on the new machine would go primarily to their elders and had been surprised by the thorough integration of their numbers into the new team. The elder bosses like Furth and Grove came to the regular meetings but usually let their young protege Goldston moderate the debate. Visitors from other labs, including Russian scientists, were routinely allowed to sit in on these candid sessions.

There were weekly clashes over experimental results, new theories, and whose experiments were scheduled for the upcoming week's "run." Everyone wanted a piece of the action.

In a given week, if the high voltage power systems did not break down, the machine operators might be able to crank out 200 attempts, or "shots," at making plasmas. Each days' worth of shots cost the lab about $20,000 worth of electricity from Public Service Electric & Gas Company, New Jersey's biggest utility. Money, however, was hardly ever spoken of by the physicists. They would run experiments on as many good plasmas as they could make – that was the limiting factor, not money.

One by one witnesses would come to the front of the conference room clutching graphs traced on transparent plastic sheets. The graphs were projected onto a wall and roundly criticized by the exacting group. The graphs showed measurements of the plasma's temperature during different experiments, graphs of the density, and statistics on the impurities present. That was the evidence. Then there were analytical graphs predicting how the plasma should behave under different conditions. It was all a very human attempt to impose an order on the turbulent, seemingly haphazard movements of the gaseous plasma – to find logic and consistency where perhaps there was none at all.

Behind the effort was an unquestioning belief that an order, a pattern, indeed a law existed that explained in mathematical terms the writhings of the plasma. Moreover, the group believed it had the collective intelligence to perceive that pattern. Once the modus operandi was cracked, the scientists presumably could bend the plasma to their will and make power-producing nuclear fusion.

There were many options for manipulating the TFTR plasma. They could change the current strength in the magnets that ringed the doughnut-shaped vessel both horizontally and vertically. They could vary the amount of hydrogen gas they puffed in. They could alter the timing of the injection of gas and the timing of the electric pulses that they sent through the belt of plasma particles. There was a seemingly infinite combination of variables to try.

Phil Efthimion was interested in employing the tried and true "ohmic" heating method, sending electricity through the ring of plasma itself so that the plasma warmed up the way the coils of a toaster do. Graeme Tait, the sharp-boned Australian, was partial to "compression" experiments, tightening up the magnetic field lines to squeeze the plasma, making it hotter because there was less space for its energy to move around in. Other experimentalists were eager to try the neutral beams. A visiting squad of Oak Ridge physicists had brought along a pellet injector that could manufacture and shoot frozen hydrogen fuel pellets deep into the vessel instead of puffing in hydrogen gas. The pellets created a greater central density of particles.

The scientists could change the interior surfaces of the vacuum vessel, covering the walls with special materials that absorbed the carbon and oxygen impurities. They could put a giant tong inside the machine called a "moveable limiter" that could actually pinch the plasma ring into a tighter circumference. There was so much to try and so little proven theory to go on. They hardly knew where to begin.

The Princeton plasmas were the most fully described plasmas in the world, and with each shot came reams of data. Altogether nearly forty different measurements of the conditions inside the vacuum vessel were being taken.

Efthimion, one of many control room habitués, was most concerned about the density of electrons in the plasma cloud. Other experimentalists kept track of ion temperatures, of impurity levels in the plasma, the amount of X-rays coming out, the infra-red rays, the visible light, and the all-important neutrons that were given off when fusion took place.

They were overloaded with data. Each shot sent more than seven megabytes of data into the computer system – equivalent to 2,500 pages of the Encyclopaedia Britannica – and they were banging out forty plasma shots a day.

At the center of this avalanche was Goldston. Using an intricate computer program called "SNAP," Goldston would work the keyboard as SNAP checked one piece of the diagnosis against others. If the information agreed, the program went on to calculate the temperature and confinement time for the plasma and compare the amount of power poured into the shot against the energy created. This was a critical comparison for designing a practical fusion reactor.

Goldston thought of SNAP analysis as an "instant" report card that took fifteen minutes to produce. With the next shot approaching, he could not waste time or be distracted. He could not stop for a soda, or a chat with his friends at other consoles. Like a seamstress in a sweatshop he was bound to his work station.

Nearby, the shift's Chief Operating Engineer, the engineer in charge, held a walkie-talkie linking him to the other engineers out at the power supply rooms and the neutral beam area. He often played the part of a NASA flight commander, for the computer screens, the countdowns to each plasma attempt, the tension, and the terminology all added to the space center atmosphere.

"Control room to Scotty!" he would shout over the walkie-talkie to the power supply engineer, mimicking characters in the Star Trek television series.

Every few months when the big machine went "down" for modifications, the physicists were even busier, plowing through the mountains of data, trying to understand what had happened. Occasionally, this world of numbers and abstractions would be abandoned for a trip out to the test hangar to see the machine firsthand instead of through the filter of computer readouts. On the rare dates that the vessel was actually opened, it was hard to resist clambering inside for a peek, just to march about where multimillion degree temperatures routinely survive. The curious were obliged to first change into white paper jumpsuits and exchange street shoes for clean white sneakers. Paper shower caps and rubber gloves completed the outfit. This comical crowd then slid into the vessel via an open portal. The engineers referred to the flock of visitors as "the Oooh-ahh! birds" because of the sounds they made as they toured their mysterious machine.

Standing upright inside the stainless steel vessel one felt like a Lilliputian trapped in a toaster oven. The floors and ceiling of the vessel were ribbed in steel while the walls were finished in neat, black carbon tiles. The moveable limiter looked like a fat tuning fork projecting from a side wall. There was room within the vessel for at least ten people to shuffle around.

A few spots on the wall that originally had not been perfectly curved – off by just a few millimeters – now showed splatter marks where the passing superhot plasma had obliterated the projections. A blue irridescence on the floor was due to burned off impurities.

One worker was assigned to survey the vessel with a geiger counter for radioactivity. Neutrons emitted from the fusion reaction displace particles in the walls, which in a reactor would cause eventual radioactive deterioration of the machine's innermost wall structure. The splattered areas of the machine – where the geiger counter started chattering – were roped off, but people were still allowed to circulate in the machine. The radioactivity was really quite low for now – a short visit inside the tokamak gave a dose equivalent to that received on a cross-country plane trip, the Princeton scientists estimated. TFTR was just beginning its experimental life. Later on, the neutron activity during plasma shots would be so intense that no one would even be allowed in the machine hangar until several days after the experiments, time for the radioactivity left in the walls to die down.

Good plasma physicists learned to be fastidious. Their plasmas demanded a "clean" machine, free of the errant atoms that might steal heat from the bottlecapful of hydrogen fuel. At Princeton, the physicists were always complaining about the "filthy" vacuum vessel even though a look inside revealed a stainless steel interior as gleaming as the inside of a brand new appliance.

"Clean" took on a new depth of meaning. It meant no air, no water vapor, none of the sticky molecules of carbon and oxygen that coat any object exposed to the worldly atmosphere. Clean meant reducing the particle population in the vacuum vessel to only 30 million particles per cubic centimeter from the 10^{19} particles in normal air. Conceptually, it was like weeding out the entire population of California so only four or five people were left to wander around.

The unwanted particles causing the most problems were invisible carbon and oxygen that stuck in layers 100 atoms deep on the vessel walls. The most the hot plasma would tolerate was a one-atom layer of "dirt." In the minds of the vacuum specialists and physicists, these layers of atomic filth were as gunky as a tub of turkey grease spilled on the bottom of an oven. Mere air pumps – even with jet engine-sized rotator blades – could not suck out the impurities. Instead the scientists turned to the same cleaning method that General Electric and Westinghouse had perfected over the years for their kitchen appliances. They used the machine's own heating capability to burn off or vaporize the unwanted atoms, just like a self-cleaning oven.

First they would heat the entire vessel to about 150 degrees centigrade. This was called "bakeout." Then "cool" inferior plasmas were produced to rub up nicely against the dirty walls. The heat of the plasmas converted the oxygen and carbon layers into carbon monoxide gas, which could then be sucked out by the air pumps. There was nothing glamorous about it, but it had to be done.

In a clean and airless environment, a cloud of plasma at least had a chance to survive. Yet many plasmas continued to protest mightily as they were squeezed tight in the magnetic fields and heated to the hilt with electricity and neutral beams. Like the plasma in PLT, those produced in the big shots on TFTR would frequently "disrupt," dumping their energy onto the walls of the vessel.

The physicists watching television screens in the control room would wince at the sudden flash of light and at the metallic boom that reverberated over the microphones from the test cell. When the plasma disrupted, it sounded as if Thor himself was taking whacks at their precious machine with a mammoth mallet.

The physicists could not explain disruptions, but it was obvious that they were dangerous to the multimillion-dollar tokamak. Heat was not the problem. The experimental plasmas were very hot but so thin that they could inflict little damage on the stainless steel. The real danger came from magnetic forces. The thunder of Thor resulted from the unleashing of tremendous magnetic forces generated by the current the plasma had been carrying. When the plasma suddenly disrupted and lost its current, that magnetic force had to go somewhere. The physicists believed that the forces reappeared on the walls of the machine and, interacting with the magnets that circled the outside of the vessel, conspired to give the impression that they were ripping the machine apart.

The clashing and groaning sounds after disruptions came from the vacuum vessel straining under the magnetic forces and then popping back into its usual doughnut shape. The new Princeton machine was buttressed to withstand considerable force, but its keepers could not be sure that a particularly bad disruption would not damage the machine. As a result, the plasma kept the scientists on the defensive, forcing them to proceed cautiously up the ladder of higher electric currents, magnetic fields, and temperatures.

In the control room it was not uncommon to hear anthropomorphic references to the adversarial plasma. It was "not behaving" or it was "unhappy" or it "liked" one heating method over another.

"To think that the plasma has some kind of active malevolence," said Furth, "that's a very tempting thing because it does have these little response mechanisms like an organism, and many of those frustrate what we're trying to do. To think that the plasma's a little malicious is not so farfetched."

He continued:

> I once made a nice personification for Lyman Spitzer that he later used in some articles, so he must have liked it. He was just working in some new, very complicated Stellarator scheme for confining plasmas, so I wrote him I thought it was very nice, but I didn't think the plasma would be intelligent enough to appreciate all the intellectual beauty of this thing . . . in other words, it wasn't going to work.

Rob Goldston thought of plasma in anthropomorphic terms as well. "The machine doesn't have the perfection and grace that plasma has," he observed.

"You have to give it a place to be and let it do its thing, and it does nice stuff. What I regret about being pushed to do things quickly, it's like a relationship with a person – you miss a lot of things if you don't take the time to understand him. It, or he, is a very mysterious being with fancy things it can do if you just give it time."

There had been time enough for exploring relationships with the plasma on earlier machines, but not for TFTR. The U.S. government had set this machine on a speedy, goal-oriented schedule, and competition from abroad added to the urgency to come within reach of breakeven. Perfection and grace were irrelevant. The machine just had to work. And soon.

With Princeton's Christmas 1982 deadline triumph, the United States began operating the world's first giant tokamak, but the rest of the pack was not far behind. Just six months later, Europe's JET chalked up its own "first plasma."

The JET lab outside of Oxford was a classy showcase of the European Economic Community's political will, and it was kept impeccably groomed for the dignitaries who streamed in to view the diplomatic phenomenon of international Big Science. The lab had a distinctly different personality from its freewheeling American cousin.

Inside, the appointments were clean and neat, creating an air of quiet formality. Everything seemed under control. The scientists dressed appropriately for a showcase, most in jackets and ties. Once in the morning and once in the afternoon, bells would chime in the hallways signaling the appearance of matrons in white uniforms pushing carts carrying tea and cookies. JET was utterly civilized.

The scientists worked at their computers under dimmed light in the cool, almost cold JET control room, which was encased in smoked glass. A thick sheet of clear, soundproof glass bisected the room, segregating the physicists from the engineers. Visiting American scientists, accustomed to shouting instructions at one another, were incredulous that the two constituencies were kept apart. It was the wish of Paul-Henri Rebut, the JET designer. The actual machine operators – the engineers – needed the quiet, he insisted, to reduce the chance of human error.

To limit other vagaries of experimentation, the scientists strictly planned each day's attempts at producing and manipulating plasmas. The engineers would program into the computers the magnitude and timing of power needs, leaving the physicists to gather the data after each "shot." There was none of the improvisational, seat-of-the-pants operation that characterized the more integrated Princeton team. If adjustments in the day's plan proved unavoidable, JET's physicists and engineers would retire to a conference room outside the control room and make an official change. The conference room also doubled as a snack bar. No food, not even a cup of coffee, was allowed in the control room, where sticky liquids might cripple a computer keyboard.

In strictly scientific terms, however, the two rival labs – Princeton and JET – were quite similar. Each coped with the same demands of the plasma. Each suffered the same frustrations and found comfort in the same small triumphs. There were power-equipment breakdowns, vacuum leaks, and computer malfunctions on both sides of the Atlantic. There were plasma instabilities and disruptions. In fact, one early experiment on JET aimed at creating tall, skinny plasmas resulted in a disruption inside the vacuum vessel so immense that it actually lifted the 100-ton vessel nearly half an inch off its moorings, dropping it with a slam that was heard throughout the vast building. According to later estimates, a force of about 300 tons had been exerted. All work was halted as Rebut ordered additional buttressing of the reactor and a correction in the computer feedback program that should have prevented the skinny plasma from tipping onto the cool vessel walls and dumping its energy.

Within days of this accident, several top people at Princeton had all the grim details. When they described the event before the weekly meeting of the team that operated the Tokamak Fusion Test Reactor, only nervous tittering swept the room, for the Princeton scientists knew they could have easily suffered such an accident themselves.

After so many years of open exchange, the Princeton–JET rivalry was no abstraction. The faces of the opposition, their personalities, intellectual strengths, and vulnerabilities were well known. So, when bad news or good news from abroad hit the lab grapevine, the scientists knew exactly who was suffering the consequences or wearing grins, and more likely than not it was a friend. At the top rungs of management, it was common for old rivals to make trans-Atlantic bets on whose machine would reach designated power levels first. The payoff usually took the form of expensive regional wines consumed at gala restaurant dinners.

This highly personalized competition was a vital force in keeping the spirit of the fusion mission alive – for individuals as well as for governments. Without a rival, a scientist facing a lifetime's battle with the plasma might give in to intellectual exhaustion. As an applied science, fusion's goals were well defined and calculated. The scientists could easily chart their own progress and measure themselves against others.

"You make it a race and almost a personal competition to give the troops a goal," said Hans-Otto Wuster, JET's West German director. "Take a flag, put it on the next hill and say, 'Now we're going here.' It's the only way to make people work sometimes inhumanly hard. Otherwise we'll just get stymied. Keep that spirit up. They see the small success. You can't give them the big success. You use every trick to keep their adrenaline flowing."

Although the cloak of secrecy was officially removed from fusion work, the pace of international competition still hinged on the flow, and control, of information. A subtle protocol emerged when it came to sharing data and insights. As allies against the plasma, the scientists shared what they knew; as rivals for prestige and patents, they controlled exactly when that information could reach their friends and how much detail would be provided.

Sometimes the younger experimentalists did not feel the rivalry as keenly as their elders, who were more accountable to government financiers. Wuster, for example, had forbidden the release of data from JET in any form other than official papers. JET's information was technically the property of the contributing European nations, and Wuster took his role as custodian with utmost seriousness. A large and intimidating man, Wuster was seldom crossed. However, the younger researchers on JET were eager to trade information with their friends working on Princeton's TFTR, if only to elicit some comment from peers before setting their analyses in print. On more than one occasion they quietly contacted Rob Goldston and Phil Efthimion at Princeton and set up an underground railroad of information. At international meetings and during individual visits, graphs and computer printouts were surreptitiously traded. A formal agreement on this sort of cooperation might have taken years to hammer out. To the young scientists, it only made practical sense. The real opponent was the plasma.

The fusion rivalry was not restricted to the two big labs. By the early 1980s, the competition to deliver temperature, confinement, and density records was extremely broad. There were nearly 300 fusion devices in the world, including seventy-three tokamaks. Japan had fifty-seven research machines. West Germany had sixteen. Each of the world's industrialized nations wanted to play the game. Australia had five devices, France had four, Argentina – the site three decades earlier of Juan Peron's delusional announcement of fusion success – had three small machines. Iran had two. World spending on fusion research was about $1.3 billion annually, by U.S. Department of Energy estimates. Although only the giants were capable of reaching the interim goal of "breakeven," many scientific milestones were waiting to be claimed along the way, and some were captured by less powerful machines.

While the giant tokamaks were still under construction or just warming up, smaller devices in the world arsenal stole some of their thunder. In 1981, a team of West German scientists at the Garching lab near Munich happened on certain conditions under which a neutral beam-heated "shot" would suddenly and clearly exhibit superior confinement of the plasma and its energy. The Germans called this enhanced confinement the "high mode" or "H-mode." Until that time other beam-heated plasmas, including those inside the Princeton Large Torus, had shown progressively worse confinement as the beam heating increased. The Princeton plasma could be made very hot but only by continually pouring on the external heating. It could not efficiently store its own fusion-generated energies. Not surprisingly, this was called the "low mode," and it would be insufficient for reaching breakeven or the ultimate fusion goal of "ignition."

The West German tokamak, called Asdex, had the special feature of a magnetic "divertor." This configuration of magnets siphoned off straying particles, keeping the edge of the plasma relatively hot. Other divertor-equipped tokamaks, including a modest one at Princeton (the Princeton Divertor Experiment), also briefly reached the H-mode. But it was a tricky plasma state to pin down and only minimally understood.

With so many fusion research devices running experiments, collaboration became a necessity because of a worldwide shortage of experienced plasma physicists. If a lab shut down a machine for repairs or modifications, scientists were frequently dispatched to other needy domestic or foreign labs. The borrowing lab paid a fee to cover the workers' salaries. When the Oak Ridge national laboratory, for example, turned off its tokamak for an upgrade, one platoon of scientists spent several months in Kharkov in the Soviet Ukraine, helping the Soviet scientists figure out what was wrong with the alignment of magnetic fields on their latest Stellarator.

International collaboration kept laboratories financially healthy in other ways, particularly by avoiding mass duplication of experiments. In 1979, General Atomic Company in San Diego, the only American private laboratory doing major fusion research, entered into an unprecedented agreement with the Japanese to share work on a tokamak.

Founded in 1956 by the industrial concern General Dynamics and then purchased in 1967 by Gulf Oil Company, General Atomic survived the late 1970s primarily through contracts signed by the Department of Energy for diversified energy research, including fusion. The tenacious director of General Atomic's fusion program was Tihiro Ohkawa, a member of the early generation of Japanese physicists who had chosen to work in the United States when Japan initially decided on a slower, basic research-oriented approach to fusion.

Ohkawa had won for General Atomic a federal contract to build an intermediate-sized tokamak capable of creating a unique, two-lobed plasma. He called the machine Doublet. To help finance a $150 million upgrade and experimental program for the machine in 1979, he negotiated a five-year collaborative agreement with the Japanese government which, under pressure from Washington, had been looking for American investments as a way of redressing a trade imbalance, Ohkawa said. Moreover, Japanese fusion leaders were desperate to train their young experimentalists on an intermediate-sized tokamak in anticipation of the giant reactor Japan had already begun building. The Department of Energy referred the Japanese to the expatriate Ohkawa, and a deal was struck.

For the stiff fee of $70 million, the Japanese bought experiment time on the General Atomic machine. The agreement called for the American physicists to alternate experimentation days on Doublet with a visiting Japanese team of twenty scientists. The two teams molding D-shaped plasmas on the Doublet tokamak produced success. At the 1982 Plasma Olympics in Baltimore, the two teams reported a record value for the pressure exerted by the reacting plasma – a value crucial to practical power generation. The collaboration continued long past the original agreement, and eventually an integrated Doublet team was producing better results than most of the world's other laboratories – including the Japanese scientists working on their nation's new giant tokamak.

One significant breakthrough in plasma confinement showed that the smaller labs, with their compact machines, were not to be summarily swept aside. At the Massachusetts Institute of Technology a handful of physicists and graduate students was still running a version of Bruno Coppi's midget tokamak called Alcator C. The vacuum vessel was so tiny that one could encircle it in an embrace.

The MIT lab was a throwback. It sat on a narrow sidestreet in an industrial neighborhood of Cambridge, Massachusetts, just past the Paradise Cafe and a sickly-sweet-smelling candy factory. Disheveled men from the detoxification center down the way wove unsteady paths in front of the red brick building with a glass door that said "MIT Plasma Fusion Center." The unlocked door opened to a stairwell. On the second floor a corridor of offices and workshops housed the fusion team. The lab director, Ron Davidson, warned visitors not to be fooled by the aggressively modest setup. The Alcator C, with its dense long-lived plasmas, put MIT right up there with the big boys and their machines – even better than the big boys because the experiment was built with just $15 million, small change compared with the reigning colossuses.

MIT was nothing if not resourceful, and its physicists reveled in the image of the underdog.

The Alcator C control room reflected the modesty and casualness of the rest of the lab. The smell of garlic and onions pervaded the room on days that pizza was the lunch of choice, consumed beside the flashing computer screens. The managers of JET would have been appalled. In addition to some newer, computerized consoles, there were outdated oscilloscopes. For years the MIT scientists preserved data off the scopes by shooting a Polaroid picture of the screen. Alcator C sat directly below the control room. A black metal spiral staircase connected the two floors. When the plasma in the machine spilled its energy in a disruption, the scientists could feel the thunder in their feet.

It was a charming, intimate place for scientists – the way things used to be. The researchers worked closely, sharing ideas, implementing what they liked, feeling in some measure as if they were in control of their working lives. They shared information quite freely with their peers at other labs and showed sparks of anger about the relatively close-mouthed ways of JET and TFTR and what they saw as the big labs' tendency to dismiss any advance that had not been recorded on their own big souped-up reactors.

Alcator C's moment of triumph came in 1983 – just days before an annual meeting of the American Physical Society. For the first time in fusion's long quest, the little machine simultaneously achieved the minimum density level and confinement time necessary for breakeven. Unfortunately, the third key to fusion success was missing, for the experiment took place at a temperature far too low for breakeven to actually occur. Yet, to have produced two of the three fusion requirements in tandem was a significant advance. It gave the scientists concrete reason to believe that reaching breakeven and ignition was possible.

Ron Parker, the MIT team leader, took great pleasure in announcing the well-timed news to hundreds of colleagues assembled in Los Angeles for the annual physicists' meeting. Princeton's Furth, approached by reporters, offered gracious comments when asked to assess the little Alcator's achievement. "This ranks as the most distinguished advancement yet in pushing ahead fusion research," he told The Associated Press news agency.

The key to the achievement was Alcator's compact, powerful set of magnets, which would be extremely difficult and expensive to duplicate in a larger machine. But a larger machine could compensate with higher temperatures, and Alcator had no neutral beams. "We pushed until the machine gave out with a big bang," recalled Martin Greenwald, who at thirty years old had been in charge of the experiment and had hastily written the official paper. In an understated fashion typical of physicists, he added: "I rode my bicycle home that evening, thinking something of some significance had happened."

It was a triumph for fusion, for the small laboratories, and for the tokamak. At MIT, where twenty-five years earlier the visiting Lev Artsimovich had shed tears of frustration, the Russian-designed tokamak continued to be vindicated.

In the 1980s, a clash of political ideology threatened to damage the unique relationship that the Soviet Union had forged with Western scientists under fusion's name.

In January 1980, Andrei Sakharov, the USSR's foremost nuclear physicist, father of its hydrogen bomb and inventor of the tokamak, was exiled without trial to the closed city of Gorky for his human rights activities and his criticism of the Soviet invasion of Afghanistan. Since his 1968 essay on free expression and nuclear disarmament, "Progress, Coexistence, and Intellectual Freedom," Sakharov had openly criticized the Soviet government's foreign policies and pressed for increased civil liberties at home, becoming the spiritual leader of the persecuted dissident movement in Moscow. In 1975, he was awarded the Nobel Prize for Peace.

Upon his banishment to Gorky, the fifty-eight-year-old physicist was stripped of his Soviet awards, including his title Hero of Socialist Labor. Tass, the official Soviet news agency, reported that Sakharov "was repeatedly warned" to stop "subversive activities against the Soviet state."

The American physics community reacted with outrage. The American Physical Society, which had an established Committee on the International Freedom of Scientists, began a letter and petition campaign on Sakharov's behalf.

Members of the international plasma physics community, most of whom were working with variations of Sakharov's original tokamak design, felt a special connection and obligation. But at the same time, the plasma physicists had a far closer and more valuable relationship with Russian counterparts than did scientists in many other branches of physics. They were in a bind. Making an issue of Sakharov might prove to be destructive. Fusion had been an open field since 1958 and no political vicissitude had yet interrupted the relaxed East–West exchange.

Many American fusion researchers, however, believed that some symbolic protest was imperative. "Our little contribution was to cut off all Princeton lab reports to the Soviet Union," Furth recalled. "We seized on a directive from the Department of Energy to cut off all unequal exchanges." In the previous year Princeton had sent 125 physics reports to the Russians and received just six in return. After the Sakharov affair erupted, the American lab reports stopped arriving in Moscow. "They were furious," said Furth. "It made an impression."

With the Soviet Union's December invasion of Afghanistan, another global political complication was thrown into the mix. President Carter swiftly ordered a grain embargo and other economic sanctions against the Soviet Union and a suspension of any new cultural exchanges. The Department of Energy issued a punitive directive as well. Scientific bodies were asked to reevaluate current exchanges and cut down to those only of clear and considerable benefit to the United States.

The Department of Energy had been conducting fusion exchanges with the Soviet Union under a ten-year agreement on cooperation in the peaceful uses of atomic energy signed in 1973 during the Nixon-Brezhnev era. Although American officials did not abrogate the agreement, they temporarily suspended visits. It was the first interruption of exchanges in the two-decade history of fusion declassification. Two years later, however, the exchanges resumed under the Reagan Administration.

The most consistent American fusion voices on behalf of Sakharov and other persecuted scientists in the Soviet Union emanated from Princeton and MIT. From Cambridge, Bruno Coppi, the transplanted Italian, spoke with passion and empathy for his imprisoned colleagues. At Princeton, Furth, the retired Mel Gottlieb, and the theorists Robert Ellis and Tom Stix kept the spirit of protest alive. Among the younger scientists, Rob Goldston felt compelled to badger every Russian physicist he met – from the highest officials to the most obscure researchers – on what he viewed as the unconscionable repression of dissidents and Jews seeking to emigrate. At the University of Texas, the renowned theorist Marshall Rosenbluth, who had written many joint papers with the Russians, lent his considerable prestige to the cause.

In their outspokenness, they were a minority. By and large the world community of plasma physicists was ambivalent about its role in the Sakharov affair. After the exchange agreements were renewed, some Americans hesitated to sign petitions for fear of being refused a visa for their next Moscow trip. European and Japanese physicists approached on the subject of Sakharov were quick to point out human suffering in other nations, including the United States, and wondered why they should single out a particular individual or country. Sakharov's plight had nothing to do with his being a physicist, they reasoned.

When Russian physicists were asked about Sakharov, they usually delivered the same pat line. "It shows that a good physicist is not necessarily a good politician," said Dmitri Rjutov, a Soviet mirror machine specialist. "He still has a position," Rjutov added. "He is still a member of the academy [of sciences] and a fellow at the Lebedev Institute of Physics."

Despite their reticence, the Europeans nevertheless made rumblings after Sakharov's exile about staying home from a forthcoming meeting of the European Physical Society in Moscow. Many individuals did, in fact, boycott the meeting.

Gentle pressure continued from the American side. In December 1981, when Sakharov went on a hunger strike to procure an exit visa for the wife of his stepson, the plasma physicists felt an obligation to speak out again. Their protest was directed to Yevgeny Velikhov, director of the USSR's nuclear power and fusion research program. A straightforward man then forty-six years old, Velikhov was a rising star in Soviet political circles and, as a plasma physicist himself, he was also a coworker of the Western fusion researchers. Furth knew him well and called him "Eugene."

With the renewal of exchanges under President Reagan, Furth found himself in Moscow for a meeting that happened to coincide with the Sakharov hunger strike. "Velikhov invited me to the ultraprivate sauna bath in the Moscow Hyatt," said Furth, referring to a posh, new Hyatt-like hotel, "and we discussed this mess. I explained to him the strong feelings we had about the Sakharov matter." Furth pointed out to Velikhov that, even with the resumption of formal exchanges, he could not compel the participation of individual American physicists who wanted to protest Sakharov's treatment.

"Velikhov prides himself on his ability to make good relations with the West," Furth explained. "From Velikhov's point of view the difficulty with Sakharov is impairing relations with the West, therefore he considers it a great nuisance and that's a more dominant consideration in his mind than whether Sakharov is right or wrong, patriotic or not patriotic."

Furth doubted that the fusion researchers' protests had much effect, "but it's one more straw for the camel's back." A week after the meeting in the sauna, coincidentally or not, Sakharov's daughter-in-law was allowed to leave the Soviet Union to join her husband in the United States. Andrei Sakharov took his meals. And the Americans began to meet their Russian fusion colleagues with less discomfort.

10
The plasma olympics

Nothing embodied the rivalry at the core of fusion research more than the biennial meetings of the International Atomic Energy Agency.

All the utopian talk of a commonly held passion for providing humanity with endless energy subsided once every two years, replaced by sheer laboratory chauvinism. It was time again for another Plasma Olympics, and the competition for impressive numbers was as keen as any Olympiad.

In 1984, the meeting in London was a classic, providing the platform for the first results from the world's two giant tokamaks, Europe's JET and Princeton's TFTR. It featured the usual joust of personalities and egos. In the era of Big Science, however, these worldwide meetings seemed to provide less intellectual ferment than in the past, some participants complained. Data were more plentiful than ideas.

Princeton's stretch drive to London symbolized the kind of zeal that infected all the big labs as they jockeyed for prestige among their peers and for continued financial support.

Technicians rushed to repair the hardware. Secretaries rushed to type up reports. Even the accountants were busy. They were converting Princeton's expense ledgers into Eurodollars so they could be compared with JET's finances. Princeton hoped to justify a spending increase by showing the Department of Energy that the lab was running a bigger overall operation than JET but on a smaller budget.

With one month to go before the 1984 meeting, the physicists at Princeton gathered for their usual Monday morning conference. Deci-

Paul-Henri Rebut, designer and director of JET, a $1 billion experiment. Courtesy JET Joint Undertaking.

"There is somehow a lack of courage from the fusion community, too afraid about the size, too afraid about the capital money."—PAUL-HENRI REBUT.

sions needed to be made on how to produce temperature, density, and confinement numbers that would impress their international colleagues. The scientists were not unmindful, too, of budget-cutting rumors out of Washington.

At the front of the room stood Dale Meade, the Monday game plan projected onto the wall behind him. The forty-five-year-old Meade was Don Grove's chief deputy on the Princeton tokamak. He stood six feet, five inches tall, a large man with the frame and drive of an athlete. He had graying hair, but his thick mustache and eyebrows were black. Another arresting feature was his deep voice. Four weeks remained before the Plasma Olympics, and Meade sounded more like a coach in a locker room than a project director at a multimillion-dollar laboratory.

Up on the wall were the experiments that would set Princeton apart: the ohmic heating plan that was Phil Efthimion's specialty, the compression experiments for which the Australian Graeme Tait was the expert, and the hoped-for neutral beam manipulations. Getting the beam guns in shape was the province of the veteran Harold Eubank, a wry Southern gentleman beloved by many at the lab for his patience, competence, and deadpan humor in the face of the chronic contrariness of the neutral beam system.

Meade's calendar also showed the schedule for each of the physicists' presentations in London. They would be pushing the big tokamak hard until two days before departure. A voice called from the back, "Did someone check the Concorde schedule?" Laughter erupted. But to analyze and type up the results, then haul along 600 copies of each research paper – the number required under IAEA rules – was no joke. There would not be time for overseas mail. The papers would go with the luggage.

Meade launched into a monologue on power needs, spilling out magnetism readings, megawatt figures, and megamperes like so many pages from a playbook. He said there were rumors filtering down from an international meeting in Canada that JET was already running experiments with high-powered deuterium fuel, having quickly changed over from ordinary hydrogen gas, and that the year-old European machine had already reached its maximum current of 3.4 Tesla in its magnets. Princeton's best after two years of work on the electrical system was 2.7 Tesla, though it had the potential to go much higher. This was embarrassing. Meade had also heard that JET was running currents through the plasma itself of about 3 megamperes. Princeton was typically running at 1.4 megamperes. The news was sobering. JET seemed to be making a bid to overtake Princeton. The Princeton team had better get moving, Meade intoned, and the neutral beams were the key to retaining prestige. JET's neutral beams were not yet installed. If TFTR's beams could be combined with new compression experiments, that would really be something out of JET's repertoire.

"We have to get sexy," observed Jim Sinnis, a deputy whose thankless job was to keep the machine in working readiness for the impatient physicists. "We have to do something JET can't. We can't beat JET with brute force. We can't arm wrestle them."

Priming the vital neutral beams in time for London was the job of the men known around the lab as "The Beam Team." At $90 million, the neutral beam injection system was the most expensive component of Princeton's big tokamak, TFTR, and also the most unreliable. The Beam Team was accustomed to trouble. It took hard work, vigilance, ingenuity, and daring to keep the system going. The worse the complications, the tougher was their determination and the greater their victory. The beams were outside the province of the physicists, who regarded the exotic plasma heating system as something of a mystery. It was best left to the engineers and technicians. For their part, the Beam Team took pride in the fact that the men with their PhD's from Princeton, Harvard, and MIT relied so heavily on them.

On any given day the neutral beam system might suffer electrical breakdowns. There could be vacuum leaks, as well, and strange magnetic fluctuations as the speeding neutral particle beam zapped the hot plasma.

The neutral beams had been developed by Lawrence Berkeley Laboratory and manufactured by the McDonnell Douglas Corporation, but the installation, maintenance, and modification was up to the Beam Team. The team's leader was Harold Eubank, but the day-to-day work on the beams fell to Mike Williams, a bearish young man with an auburn beard and an intensity that he wore like armor. Williams was smart, resourceful, and utterly devoted to Eubank. Like Eubank, he never offered excuses. His job was to make the beams work. Period.

The Beam Team was very young. Most of the dozen men were in their twenties and early thirties. The job required physical stamina, for beam installation was relegated to late shifts and weekends. With the London conference deadline pressing down on the lab, Meade opened up time for the beam boys to work during the day. So they would be working night and day.

The Beam Team labored in the beam control area on the floor directly above the tokamak test hangar. The computer cabinets and data screens were there, side by side with the electronics. Efforts to link the beam monitors to the main control room were far behind schedule. So the Beam Team sat amid the computer cabinets with the noisy whine of the giant airconditioning system and other machinery drowning out any conversation from more than eight feet away. The noise itself was exhausting. So were the sixty concrete stairs between the test hangar and the beam control floor. Anytime they wanted to visit the men's room, grab some food or check the beam connections into the tokamak, it was a 120-step chore. Out at the far end of the tunnel, separated by noise and mission from the rest of the lab's workers, they felt like a frontier gang, cowboys on the range. They even dressed the part, favoring blue jeans and flannel shirts. Many wore beards. They shared with Williams a quality of stubbornness and resourcefulness. And they knew the lab both counted on them and took them for granted.

With two and a half weeks left until London, electronic and mechanical breakdowns were piling up. No sooner did Mike Williams get word from his teammates one afternoon that a balky modulator-regulator had been fixed than his computer screen suddenly showed a power failure on one of the three electrical sources feeding the beams. It might just be a sixty-cent fuse, Williams thought, but to get at it they would have to shut down the tokamak experiments for the remainder of the day. True to form, Williams acted on a hunch. He ordered the team to override an electrical safety system and let the beam shots proceed unguarded. Eubank was kept in the dark, but the gamble worked.

Such risks would have been unthinkable at JET or at Japan's lab, but the maverick Beam Team took them in stride. There would be beam-heated plasmas for London after all. That's what really mattered.

The pungent smell of sweat hung over Princeton's main control room. Graeme Tait, his angular face unshaven, was bent over a computer console. Rich Hawrluk stood behind him, every now and then pushing back a plait of matted hair. Dale Meade, the black-mustachioed coach, stood by, too, his skin a pasty white. It had been a wearying week with no time for vanity. This was it, the last day of experiments before the flight to London.

The control room was packed with fifty people. Every console was manned, every cabinet surface covered with printouts and logbooks, soda cans, coffee cups, and candy bar wrappers. Heat from bodies and computers raised the room temperature to tropical levels. Shirts were damp and unbuttoned, the sleeves rolled up.

Don Grove said aloud that everything that could go wrong had that day. A simple circuit fuse, a little larger than those found in a household fuse box, had come loose and decided to cook. The local fire department had been at the lab at 6 a.m. because the machine operators had smelled the insulation frying. Then a magnet power supply failed in the test hangar and had to be fixed. None of this had anything to do with plasma physics. It was just the sort of niggling mechanical bad luck they could not afford.

Watching the whole debacle with a bit of smugness was a young Japanese visitor named Mitsuru Kikuchi. He had been on exchange at Princeton for the past several months, gaining experience he could use when his own country's giant tokamak started running in the spring. As the Princeton scientists lurched from one crisis to another, Kikuchi reflected that this would never happen on the new Japanese machine. It had been built by Japanese industry at about twice the cost of the Princeton machine – expensive, yes, but absolutely guaranteed by the manufacturers.

Here, half the day or more was being spent on troubleshooting. If a part failed on a Japanese machine, the name of the company responsible would be announced widely to the public, Kikuchi explained. The company would lose its reputation; its "shin-yo," or trustworthiness, would be damaged. Moreover, it would be up to the manufacturer to fix any problems, not the lab engineers.

When Princeton's Chief Engineer finally got word that the tokamak and support systems were ready, he yelled over to Tait. The Australian checked the instructions they had programmed into the computers for this particular "shot" of plasma, the timing of the neutral beams, the amount of deuterium injection, and the strength of the magnetic fields. The engineer began the countdown and Tait pressed the timing clock button. At 5:30 p.m. the neutral beams pulsed 1.1 million watts worth of energetic neutral particles into the hot plasma.

Then the plasma rebelled. For hours the men got nothing but disruptions as the plasma threw off the excess power from the neutral beams. A supply of pizzas and hoagies arrived and disappeared. Finally, around 11 p.m. things started to click. Four shots in a row were good. The last one, just before midnight, had 1.25 million watts of beam power into a plasma that carried 1.4 megamperes of current – impressive numbers for London.

It was time to quit the control room and head for the typewriters. The Princeton team would leave for London in two days. It had been a frenzied, exhausting home stretch. The rhythm of competition had caught them up again, creating a ripple of excitement in a task that would consume a lifetime.

More than 600 physicists registered for the London meeting, the tenth International Atomic Energy Agency Conference on Plasma Physics and Controlled Fusion Research. Among them were 126 from the United States, 74 from Japan, 53 from the United Kingdom, and 59 from the separate European entity of JET. The Soviet Union registered 25 delegates and the People's Republic of China 10. This was a special Plasma Olympics, for it would include the first experimental results from the two giant tokamaks and a celebration in Oxford where everyone was invited to inspect JET and dine on roast pig.

Assembling in the convention's conference hall, the scientists first heard Yevgeny Velikhov, chief of the Soviet delegation, deliver a memorial lecture dedicated to Lev Artsimovich, who died in 1973. Velikhov pointed out that the modern tokamak work of his assembled colleagues had confirmed many of Artsimovich's ideas and urged continued international cooperation "for our common cause of supplying power to humanity. If we want to reach something real this century," he said, "it will be possible only on the basis of tokamaks."

Then Paul-Henri Rebut took the podium and reported JET's first modest results. Phil Efthimion followed with the TFTR numbers pulled from the finish-line sprint. Martin Greenwald from MIT then gave the latest Alcator C attempts at improved heating while still maintaining the density and confinement achievements. A physicist from General Atomic in San Diego added to the data pile with a report on its Doublet tokamak.

There was no tokamak report from the Soviet Union in the same league of these four. The Russians seemed to be fading from the picture. Where once they had dominated the tokamak hardware race with their initial expertise, their ambitious plans to build a giant T-20 tokamak with superconducting magnets had been abandoned, apparently mired in budget problems. Their fallback was a more modest superconductor called T-15, but it had been hung up for years with construction problems. That left them without a modern tokamak. The venerable T-10 had set the United States scrambling in the 1970s to build a tokamak that could beat the Russians to the prize of breakeven, but T-10 had proven to be an empty threat. Identical to the Princeton Large Torus but without neutral beams, it was a machine of limited capabilities run by the same outdated computer that the British had brought to Moscow in 1969 to confirm Artsimovich's tokamak results. With paper and pencil the Russian theorists were still perhaps the world's most formidable. But the complication and expense of Big Science had put the Russians at a disadvantage.

The reports from the ultramodern JET, TFTR, Alcator C, and Doublet emphasized advances in what the scientists called "parameters," the characteristics of the plasma. All improvements had been arrived at by "brute force," by applying more powerful heating methods and magnetic systems or by using larger machines. The reports to the international audience did not contain advances in the theory of plasma behavior.

During the question period following the Doublet report, Harold Furth rose dramatically from his chair in the audience.

"This is not in the nature of a question but a comment," said Furth. Something in his voice created a hush in the audience. "Sixteen years ago," he began, "the Kurchatov Institute handed us this present, the tokamak, but it was not possible physically to explain why it worked so well. Today it is the same question. We still do not understand physically why it works better some times than others."

What are the mechanisms that cause the plasma to lose its energy? Furth wanted to know. What was the elusive key? The scientists had been reporting slightly better results, but why? "I would hope at this meeting there would be not only great advances in parameters but in understanding," he emphasized. "I hope the pieces of the puzzle will be put together. One week from now we will have a summary session. If we are lucky," he said with an edge of sarcasm, "by then we will understand tokamaks."

It was vintage Furth. He sat down, and the room was abuzz. Here was the spiritual leader of the American program apparently chastising his peers for failing to concentrate on understanding the nature of plasma. Was not Furth ultimately responsible for pushing the Princeton program to reach for the big numbers in temperature, density, and confinement, to crank up the neutral beams and see if they would work? Which lab was more intent on the "parameter" race than Princeton?

At that moment it had been the scientist in Furth winning out over the pragmatic laboratory leader. The question of brute force versus insight was an old one for him, dating back to the protest paper he wrote with Dick Post in the early 1960s.

Researchers had stumbled on ways of operating a handful of the world's tokamaks to intermittently produce higher confinement times. They called this magical region "the high mode" or sometimes "the z-mode" but as far as Furth was concerned it was "all superstition, mythology," he said later. "Like people in the Middle Ages could explain lightning and thunderstorms as the work of the Gods, but meanwhile they did practical things to protect their crops." He wanted to encourage people to think about why the high mode appeared and disappeared. He called it a "pet peeve" but it was really a deeply felt instinct as a scientist.

Ironically, Furth's appeal for a renewed emphasis on understanding the plasma might also have been a very practical, political suggestion. There was talk in Washington – in particular from President Ronald Reagan's science adviser, George Keyworth – that maybe the government should get out of the business of financing applied science. That should be industry's job. Government's job was to finance the basic, speculative research in which profit-making industry could not afford to invest.

Furth was recognizing a need to do both basic and applied research simultaneously. He wanted to push ahead with the big machines but not at the expense of abandoning the more difficult challenge of puzzling out the behavior of fusing plasmas. The world was wrapped up in a hardware competition. It should also be wrapped up in a competition of ideas.

The Japanese were proposing to build a $2 billion "Engineering Test Reactor." The Europeans had a similar design on paper to succeed JET. Princeton had proposed a $1 billion ignition project, to replace the giant tokamak they were now running. Big Science had fueled a hardware competition of enormous proportions, and the rivalry showed no signs of letting up. But it was difficult to know whether the fusion race was headed in the right direction.

11
Different directions

The Lawrence Livermore National Laboratory occupies a fenced-in square mile of flat land in the Livermore Valley of northern California, where the land is baked to a brown and barren crust in the summer sun but grows lush in the cool months. Livermore's history is evident in its eclectic architecture. Wooden barracks left over from the site's use as a World War II naval air force station still stand among plain brick structures, metal trailers, and tall, modern glass offices housing the latest physics research machines. All visitors to the complex, whether destined for the unclassified labs or the secrecy-shrouded weapons research buildings, go through a formal check-in procedure at a gatehouse. There, they are photographed and given badges and daily passes.

Livermore is a lab for the nuclear age. Within its confines scientists pursue, without inhibition, a wide range of interconnected atomic questions. How can better bombs and powerplants be made through the fission and fusion of atoms? How can fissile fuels be created? What should be done to reduce or store the radioactive waste produced by atomic reactors? What are the effects of radiation on living things? How should a nation deal with nuclear accidents? Could a spaced-based defense shield using laser and particle beam weapons be created to fulfill Ronald Reagan's dream of a Strategic Defense Initiative, or "Star Wars" program?

In one far corner of the Livermore grounds stands a modern building of dark brown smoked glass – to discourage enemy binoculars it is said. There is no special fence around the building, but its entrance leads into a kind of security closet. A visitor is required to shut the door behind him, identify himself, and stand under the scrutiny of guards using remote-control cameras. Once the visitor is approved, the first door locks and a second door, leading into the building, opens. Visitors without prior security clearances issued by the government are accompanied at all times, even into the restrooms. Sandwich-board signs are propped up in the hallways to alert the scientists and clerks within. "Caution," they warn. "Uncleared Visitor Under Escort" and "Visitor. Unclassified Discussions Only."

Inside this carefully guarded building, plasma physicists sworn to secrecy ask their own atomic question: Can the heat of powerful lasers be used to create controlled nuclear fusion?

The declassification of fusion research at the 1958 Atoms for Peace conference did not apply to every fusion scientist for all time. Collaboration with the Soviet Union, the Japanese, and others on magnetic fusion devices, such as tokamaks, mirror machines, and Stellarators, was deemed acceptable. But the Livermore scientists in the brown glass building experiment on a different fusion reactor scheme, one based so directly on hydrogen bomb physics that it cannot be openly discussed, at least not by the U.S. government and its employees.

With the advent of high-powered laser beams in the 1970s, the Livermore researchers began bombarding minuscule pellets of hydrogen with pulses of focused energy, setting off what amounted to miniature hydrogen bombs – explosions so small that they could be contained in an instant within a room-sized chamber. If the pellets could be exploded rapidly and regularly and with an efficient use of power, the energy might be captured and transformed into steam to run a power plant. The scheme was called laser fusion.

What interested the American military and its censors was that these miniature laser fusion explosions could simulate certain conditions produced by a genuine hydrogen bomb. A laser fusion miniblast would generate ultrahigh pressures and temperatures and release the same kinds of atomic particles. These basic experiments could be useful in testing, at relatively low cost, the effects of nuclear explosions on various materials used in satellites, warheads, and other military equipment. Moreover, perfecting laser fusion pellets would have applications

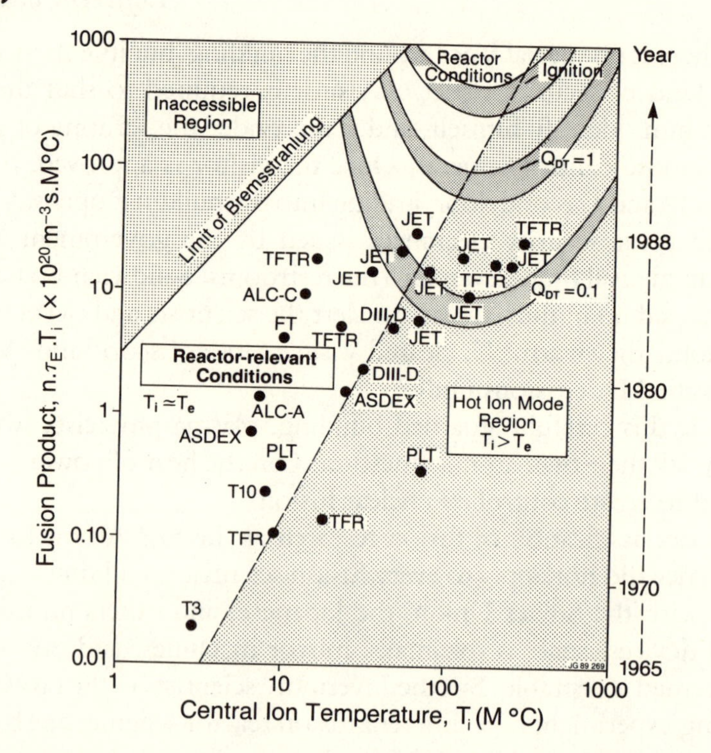

This graph shows the progress over the years of various fusion test machines in moving toward the goal of reactor conditions (upper right). Horizontal axis represents plasma temperature in millions of degrees centigrade. Vertical axis represents plasma density and confinement time. The graph is in logarithmic scale. Thus, increments at the top are vastly greater and more difficult to achieve than those at the bottom. Courtesy JET Joint Undertaking.

to the design of real bombs. Eventually, if made powerful enough, laboratory laser fusion could actually replace altogether some underground nuclear tests.

Fusion had been sold to the world on the utopian promise of clean, safe, inexhaustible energy. It was a technology of benign intentions and humanitarian goals carried out openly among the world's scientists. This was fusion's pristine image. But in some quarters this pose was deemed inhibiting and naive. There were directions that fusion research could take – more successful directions, some believed – if the fraternity was not so sensitive to public image and insistent on segregating fusion from fission and energy research from weapons technology.

It was easy to see why Livermore, the nation's premier weapons lab, became the crucible for some of modern fusion's most difficult image problems, the place where hard questions converge and are faced head on. Peaceful fusion and destructive fusion live under the same roof, and some of fusion's adherents felt it would be folly to pretend otherwise. Questions of public image and self-image could not be avoided. Any individual working at Livermore continually faces the question posed by relatives, friends, and acquaintances – is he working on weapons? There were other image issues faced by the world's fusion community – chief among them the question of fusion's limited but inescapable radioactive hazard – but Livermore scientists bore the brunt of the really tough issues.

A smaller laser fusion project was initiated at Livermore in the early 1970s with the same high expectations and secrecy that had accompanied the start of Spitzer's Matterhorn Project at Princeton two decades earlier. The first lasers, however, were not powerful enough to produce the combination of high temperature and density needed for a self-sustaining reaction. By 1980, a $176 million project to build the world's most powerful laser, called Nova, was under way at Livermore. The program was financed not by the energy research division of the Department of Energy but rather by the military applications division.

This funding enclave provided the laser scientists with a valuable constituency: the military establishment and its political benefactors in Congress. The Senate and House armed services committees, for example, became strong and steady supporters of laser fusion, which was viewed as a weapons program. In fact, the scientists had rediscovered fusion's oldest constituency, for the original Atomic Energy Commission, from which the Department of Energy had evolved, added Spitzer's controlled fusion research to the Matterhorn Project with an eye toward fusion's unknown military applications.

Fusion's kinship with weaponry eventually came full circle. After years of open research, East–West collaboration, and the distancing of the science from its nuclear bomb origins, the fusion community found in the 1970s and 1980s that the new laser technology was pulling some of its members back behind the classified fences from which Spitzer and others had sought release.

On the surface, at least, the lure of laser fusion is evident. Its key is the laser's powerful snap, its concentration of energy in time and space. In the early years, the scientists pursued what they called "direct drive" fusion, in which several laser beams were aimed directly at a tiny plastic sphere containing deuterium and tritium gas. The intense heat is absorbed at the sphere's surface causing it to boil off and rocket outwards, driving the contents inward, creating the hot, dense plasma conditions necessary for fusion. The laser could hit the plastic sphere with up to 100 trillion watts of power in a burst that lasted 1 billionth of a second.

The resulting explosion was relatively small because of the tiny amount of fuel involved in the laboratory experiment. Instead of the several grams of hydrogen isotopes used in nuclear bombs, the fusion physicists were dealing with mere specks of deuterium and tritium, about 20 millionths of a gram. The fuel target was only several hundredths of an inch across, smaller than a tiny seed.

But the direct drive approach left a negative balance sheet. The laser energy needed to trigger the reaction was extraordinary and did not produce as much fusion energy as predicted. Under real world conditions, the scheme was terribly demanding. To produce excess energy, the fuel target had to be compressed to a size about thirty to forty times smaller than its initial size without becoming badly asymmetric and breaking up in the process. This required having an exceedingly smooth plastic sphere – smooth, that is, down to the atomic level of the surface, a daunting prospect. And the laser beams would have to exert enormous pressures on the sphere in an absolutely uniform manner, not differing by more than a few percent. Sixty individual beams (and perhaps as many as 100) would have to be perfectly focused. Any hope to pull off such a feat with contemporary equipment proved to be unrealistic.

Eventually, the Livermore scientists turned most of their efforts to a two-step approach called "indirect drive," in which a fuel capsule is placed in a radiation case of gold or lead. The Livermore scientists then "somehow" convert the laser beams into soft or low-energy X-rays directed into the case. And "somehow" the X-rays create a kind of shock wave that compresses the fuel with uniform pressure, a great advantage over the earlier method.

"It's all the 'somehows' that are classified," Erik Storm, director of

Livermore's laser fusion program, explained. "The reason they're classified is because there is an obvious connection between X-ray driven laser fusion power and X-ray driven weapons, and so we don't want the details of that to come out."

"I want to make clear," Storm added, "we cannot make weapons with laser fusion. We can't test nuclear weapons. But on a very, very, very small scale you can duplicate part of the conditions that exist inside thermonuclear weapons. You can do physics experiments that are of interest to the weapons community."

The only other option for the military was to test materials by placing them next to a nuclear bomb blast at the Nevada Test Site. This would obliterate the materials and the diagnostic devices within millionths of a second after the explosion – a very expensive way to gather data.

The scientists in laser fusion were by and large much younger than the magnetic fusion physicists. The laser was a new technology, after all, that came into vogue just as they were completing graduate school in the 1970s. Some were attracted to laser fusion because of the chance to work on new, high-technology problems. Others shared the utopian endless-energy dream of the magnetic fusion crusaders. They did not consider themselves to be weapons scientists. They did not make bombs. It was important to them that people realize this.

Erik Storm, a blue-eyed native of Norway, said he had become an American citizen as a prerequisite for taking a classified laser fusion job at Livermore in 1974. Ten years later, at age forty, he was the program director, delivering laser fusion's message in a smooth and rapid California idiom.

He described the work at Livermore with ellipses, and he expressed a frustration common to the residents of the smoked-glass building who were bound to secrecy. The lab's atmosphere was electric with excitement, and the researchers bore Cheshire Cat smiles, but they could only brag mysteriously about their work to people without security clearances. They were on the right track and making great progress, they said, but to reach the energy level necessary for a laser fusion powerplant they would need a laser that could deliver fifty times more energy than

Nova. It would probably have a price tag of about $1 billion. Congress was simply not ready for that. But experiments would continue, nevertheless.

Although laser fusion replicated in miniature the workings of the hydrogen bomb, Storm preferred to liken the tiny laser fusion explosions to the birth of a star when he spoke to the public. It was a perfectly valid analogy. Yet the researchers could not escape the fact that their idealistic energy work was piggybacked onto a national defense program and that they had gone "inside," the slang expression for working within the figurative security barriers surrounding classified research.

To generalize about the effect this weapons connection has on the attitudes of the laser fusion researchers would be a mistake. Their views range from full support for the connection, as a valued financial underpinning for fusion science, to persistent uneasiness.

One can certainly generalize, however, about their attitude toward secrecy. They despised it. There is no justification for applying it to their work, they contended.

The sentiment expressed by Mike Campbell, a top laser researcher at Livermore, was typical. Campbell had studied plasma physics at Princeton's graduate school but had passed up magnetic fusion in favor of the "more challenging" lasers. By age thirty-three, he had the title associate program leader, and in blue jeans, sneakers, and dark brown beard, organized both the experiments and the group's afternoon softball games. Campbell was contemptuous of the secrecy process. "I look at classification as set by people who don't know what should be classified and unclassified," he said. "Although it's a marginal connection (to weapons), it gets classified because a nontechnical bureaucrat in security tends not to take risks. They're extremely conservative."

Campbell was irked as well that his peers in the fusion community might assume that laser fusion's silence meant it was floundering. "What you find is that people expect that when they hear nothing, you've had difficulties and problems," he said. "Everyone expects that if there's a great success it finds its way out no matter what it is." Consequently, he said, laser fusion was always described as being at least a decade behind magnetic fusion. That, Campbell maintained, is "far from the truth. I think in certain elements it is ahead of magnetic fusion."

Campbell told of an article he submitted to *Nuclear Fusion,* the premier international journal of the field, "to help show that laser fusion is viable." It was rejected by the referees as "cryptically described" and "obscure." In order to publish, the referees said, more information on the fuel pellet target was required. Campbell could not comply. "What makes laser fusion classified is the target and the physics associated with it," he lamented.

Classification also threw barriers across career paths. It was difficult to make a name for oneself. Sometimes, said Campbell, a couple of laser "shots" in an experimental sequence would be directed at dummy targets instead of actual fuel assemblies. It was all done as a courtesy to the researchers, providing some unclassified work to write up for publication.

Inside the fence, however, the pure challenges of science sustained many. The laser researchers, like their fusion counterparts on the outside, got an intellectual charge out of attacking complex problems requiring futuristic technological solutions. The Livermore researchers, for example, developed a camera that could photograph the instantaneous heating effects inside their vacuum chamber. It took a new frame every 50 trillionths of a second. "Applied science is moving to femtoseconds and submicron, ultrafast and ultrasmall," Campbell said with an edge of bravado. "By default you get exposed to that technology when you're working on the laser fusion program."

The Livermore laser people were permitted to speak freely only with their counterparts at the Los Alamos and Sandia national labs in New Mexico who also conducted secret fusion experiments and with colleagues in Britain, where laser fusion was classified. Contacts with the British fell under the same security agreement of the 1950s that governed Spitzer's contacts about his secret Project Matterhorn. The Soviet Union had a secret laser program as well.

But while Campbell and his colleagues were being patriotically tight-lipped, there was a most stinging development on the international fusion circuit. In the 1980s, the Japanese, with no system for classification and no weapons industry, began publishing papers about advanced laser fusion techniques, establishing an unparalleled reputation in the field. They openly discussed energy delivery techniques and target designs, including their "Osaka cannonball," a hollow ball to house the fuel pellet. In an irksome affront to the U.S. laser fusion researchers, American physics journals routinely published the Japanese papers.

Paul Drake, a young researcher who switched from open magnetic fusion work at Livermore to classified laser fusion work there, was infuriated by the international inequity. "The Japanese are publishing this stuff like mad," he fumed. "Since we've been at it longer, we know things that they don't. We feel that over the years we've accomplished some significant and worthwhile science that we ought to get credit for in the world, and here the Japanese are off publishing it, and there's no action out of Washington, and that's just real stupid."

John Lindl, a target design specialist at Livermore, was equally perturbed about the context of Japan's progress. "You could follow their learning curve," he said. "We could have saved them five years of work if we could have talked to them. It's a clear case where classification has hindered international cooperation. It's a big enough problem that you really hate to see resources not being optimally utilized."

Erik Storm, who as program leader was extremely careful with his words, found classification led to "very Kafkaesque" situations, especially when he was asked about the Japanese reports. "To even say they are saying things we cannot say is breaking classification," he said with an ironic laugh.

When, in 1986, a National Academy of Sciences committee reviewed the laser fusion program for the president's Office of Science and Technology Policy, it recommended a relaxation of the secrecy – and it did not tiptoe around the subject as Storm felt obliged to do. The prestigious committee wrote:

> It hurts the morale of imaginative scientists who are unable to take credit for their creative work, and often must endure the vexation of seeing nearly identical work published in the open literature, usually some years later, by workers in Japan, Europe, or the Soviet Union. Classification impedes progress by restricting the flow of information, and does not allow all [laser fusion] work to benefit from open scientific scrutiny.

Still, no significant classification changes followed the report. Until they could talk openly, the laser fusion researchers were bound to endure misunderstanding, resentment, and skepticism. They attended the International Atomic Energy Agency "Plasma Olympics" but the scant laser fusion papers were buried under the tonnage of open magnetic fusion data. At the 1986 Olympics, Storm tried to bring his colleagues up to date on American laser progress, but his audience was accustomed to more frankness. Mike Zarnstorff, a Princeton tokamak researcher, characterized Storm's talk as "strange."

"He said, 'We've done really great things, but I can't tell you and I really feel bad about that'," Zarnstorff recalled. "It was really galling."

Storm had to sit on his news for another two years.

There had been rumors in the open fusion community that the Livermore laser group was testing its fuel casings by using the forces from underground nuclear bomb tests in Nevada. In the laboratory, research had been limited by the fact that the scientists had "only" 100 trillion watts of laser power they could aim at their fuel capsules. By using the energy from a nuclear bomb explosion, the Livermore people would be able to discover exactly how much energy was necessary for power-producing laser fusion, what kind of capsules to use under high energy conditions, and whether their approach could ever be feasible as a power plant.

In the fall of 1988, the U.S. Department of Energy relaxed classification enough to shed a sliver of light on the laser program's nuclear connection. Under a program called Halite/Centurion, energy from underground nuclear explosions at the Nevada Test Site had indeed been used to implode fusion capsules and study how to design an efficient fusion target. "Experiments conducted by this means have allowed demonstration of excellent performance, putting to rest fundamental questions about basic feasibility . . . ," said the Energy Department.[2] As a result, the Energy Department was studying whether a new laser could be built "that would be capable of achieving fusion routinely and generating 1 billion joules of energy, the energy equivalent to one quarter ton of TNT." A laser capable of delivering up to 10 million joules of energy was now thought to be necessary for this laboratory demonstration.

The Nevada bomb tests had shown that with "a nasty match" to ignite the fuel, as Storm put it, fusion in a capsule could work. The overriding question was whether the United States could afford the $1 billion price of such a match.

The science of plasma physics and controlled nuclear fusion had always shared a boundary with weapons research, a boundary the fusion scientists tried not to violate. But when, in March 1983, President Ronald Reagan introduced his spaced-based Strategic Defense Initiative (SDI), the fusion community grew nervous.

The defensive shield Reagan sought to build involved nuclear-pumped laser weapons shooting through the cold plasma of space. It was a task for which the plasma physicists' expertise in nuclear theory, diagnostic work, and computer simulations could be relevant. The man who had introduced Reagan to the Star Wars concept was none other than Edward Teller, father of the hydrogen bomb and one of the founders of controlled fusion research. As SDI's chief prophet, Teller lobbied the White House, the military, and the public in favor of deploying the system as soon as possible to combat the Soviet Union's presumed weapons advances.

Although they had dedicated their careers to the pursuit of a peaceful use for fusion, the fusion community's members found that financing for their reactor research was threatened by the overwhelming $26 billion Star Wars program, one of the most expensive engineering schemes ever conceived. Scientists from MIT to Princeton to Livermore worried that making a livelihood in reactor research would become untenable and that, to continue working in plasma physics, one might have to switch to weapons-related studies.

The debate raged within the nation's general physics community about the morality of contributing to SDI and whether such a system could ever be made to work. The plasma physicists were touched personally by these questions. Discussing the hypothetical situation in which they had to choose between Star Wars work and leaving the field, some vowed never to work on weapons. Others maintained that since the Soviet Union was undoubtedly pursuing Star Wars-type technology, there was no reason for Western scientists to refuse to do it. And besides, it was interesting physics.

At Livermore, of course, these hard choices became concrete realities. Livermore was primarily a weapons research lab with an annual budget of more than $900 million. By 1987, Livermore had received direct Star Wars contracts of about $180 million, according to lab records. As the unclassified magnetic fusion program – the mirror machine work – shrank at Livermore, engineers and technicians, who were shared by all projects, were transferred en masse to other parts of the lab where SDI money was giving life to new devices. Livermore's magnetic fusion physicists, assigned more permanently to particular programs, began a voluntary trickle into Star Wars work, specifically to a project called the Free Electron Laser. It was a laser beam of intense, short pulses that might eventually work well for heating magnetic fusion plasmas – as well as knocking out enemy missiles from space. After ten years in mirror machines, William C. Turner was among those who made the switch. He did so because he thought the work was "fresh" and "more interesting," he said, plus the strong SDI funding was creating the opportunity for many experiments. The source of his research money bothered him "less than I thought it might," he said, "because I see other uses for the technologies. I guess it was a realization that a lot of the money for research does come from the Department of Defense."

The U.S. plasma physicists were not the only scientists to be presented with Star Wars opportunities. In 1986, Britain's financially struggling Culham lab won a $4.3 million contract from the U.S. Strategic Defense Initiative Office to develop a system for producing neutral particle beams. The work was to be done by Culham's particle beam group, which for a decade had been developing high-power particle sources for fusion work.

In any case, the threat of having to choose between reactor research and Star Wars proved more fear than fact for most plasma physicists. "The fear at the outset was that trained staff would go to SDI," Don Cook, head of the particle beam fusion program at Sandia national laboratory, said in early 1987. "A couple of people have transferred but not a large number." Still, Cook said, he expected that within five years Sandia's fusion program and the Star Wars program would be conducting mutual projects.

If financing for magnetic fusion and laser fusion held steady, the plasma physicists might not be faced with an invitation to cross the boundary into direct weapons research. But amid the general budget-cutting enthusiasm of the Reagan era, the fusion program appeared vulnerable, for its product would not be ready until perhaps the second decade of the twenty-first century, if then.

The Livermore lab was home to another scheme for improving fusion reactors that did not mesh well with the community's image of purity. Indeed, the fusion fraternity outside the fence rarely spoke of the fusion-fission hybrid reactor, or fusion breeder, for fear that it might sully the entire program.

Livermore was not the least bit reluctant in touting the hybrid, however. Researchers examining market forces and looking for an early application of nuclear fusion had long since hit upon the idea of using a fusion reactor to produce fuel for fission reactors. The peripatetic Edward Teller, who helped found Livermore, was enthusiastic about the idea. "Combining fusion and fission is a natural marriage," he wrote in his 1981 volume "Fusion." "Fission reactors are long on energy and short on fuel supply. Fusion reactors are short on energy, but will be long on neutrons produced, which can be turned into fuel. A fusion-fission hybrid could make a most useful contribution to the energy supply before the end of this century."

In the hybrid scheme, a blanket of uranium 238 or thorium would absorb the abundant neutrons ejected from a central fusion reaction and be converted to the fission fuel plutonium or uranium 233. It was predicted that one fusion hybrid could breed enough fuel to supply between five and thirty fission reactors, depending on the fission reactors' size. Everyone knew, as well, that plutonium was necessary for the production of bombs, but this aspect of the breeder was downplayed. The idea would be to create fission fuel for power plants more cheaply than mining increasingly scarce and expensive uranium. There was another attraction, too. The device did not have to perform as efficiently as a "pure" fusion reactor. Temperature, density, and confinement constraints were not as critical when the fusion reactor's main purpose was the prolific production of neutrons, not the economic production of electricity.

Despite interest in some circles, the breeder potential of fusion had always been minimized in the United States, scientists explained, because of the bad public reaction it might generate. Construction of new fission plants came to a halt in the 1980s, accompanied by falling oil prices and widespread public concern over plant safety and cost overruns. The demand on uranium resources slackened.

Still, some scientists predicted that the hybrid would be the first and natural application of the nuclear fusion reactor. At Livermore, the main U.S. government lab receiving funding for hybrid schemes, a study was published in 1984 showing that without some kind of breeder, the United States would run out of uranium fuel sometime between the years 2025 and 2050. Ken Fowler, director of Livermore's magnetic fusion program, told the American Nuclear Society in 1985, "Assuming that fission does make a comeback, breeding is surely the earliest application of fusion with a clear economic incentive." But Claire E. Max, author of the Livermore uranium study, wrote that "An unspoken reason for the fusion community's previous lack of vigorous support for the hybrid fusion reactor is a reluctance to 'taint' fusion by association with fission power, currently low in public esteem."[3]

This reluctance about hybrids tended to dominate the American fusion community, but the Soviet Union embraced the hybrid scheme. The Russians even declared a hybrid to be the immediate goal of their fusion program. In 1984, in his opening address at the Plasma Olympics in London, Yevgeny Velikhov described a Russian hybrid under design that would produce "a substantial amount of electricity and plutonium."

Hybrid advocates like Teller envisioned it only as a "bridge" from the troubled and finite fission power age to the age of "pure" fusion. "Fusion breeders could set the stage for second-generation fission reactors by providing an adequate fuel supply," Fowler told the American Nuclear Society. "Then, because of the environmental advantages of pure fusion reactors in eliminating long-term waste storage, fusion would gradually replace fission altogether. This is an old vision, but one that deserves reexamination in light of today's knowledge."

The utility companies liked the hybrid because it was a way of testing and perfecting the fusion reactor without having an entire power grid relying upon it. "It could operate on the side as a fuel factory and at the same time get the bugs out of a fusion reactor," said Betty Jensen, a research and development manager at Public Service Electric and Gas of New Jersey. "It's an excellent way of getting the fusion reactor on line . . . a practical way of bringing in fusion."

The Soviet Union saw it that way and it took the hybrids seriously, but the American government's hybrid effort dwindled to just a handful of researchers.

Ralph Moir, a nuclear engineer who headed the Livermore hybrid group, was understandably embittered by the whole process. The scientists who began the nation's fusion program, Moir said – people like Dick Post – "were very happy to accept the money even though they wouldn't accept breeding as the mission. They said, 'We're going to do more noble things.'"

The fusion-fission hybrid had become, to them, "a bad word, a bad connotation," Moir related. The truth was, he said, that "commercial nuclear power and nuclear weapons have connections in technology or even just ideas but how many fusion people want to contribute to weapons making? I mean, that's not nice. We would like to say we're contributing to making daisies all over the world. Nice. Benign. But damn it, the world isn't that way."

Although there were pitfalls in the politics of nuclear energy development, the hybrid was modest by comparison. The most yawning hole was that of radioactive fuel and waste.

Fusion had always been touted as clean energy. Yet it was only relatively clean, relative to fission, that is. So long as the fusion researchers were conducting basic experiments using hydrogen and deuterium as fuel, there were no radioactive hazards. But the drive toward breakeven and the design of a demonstration reactor in the 1990s and beyond would demand the use of tritium, a radioactive form of hydrogen. Tritium was simply a more efficient fuel. It could not be ignored.

The fact that fusion would inevitably involve some radioactivity had to be handled delicately by the scientists. Fusion had no strong public image, and already it was frequently confused with fission, which of course had an abysmal image after the nuclear power plant accidents at Three-Mile Island in Pennsylvania and Chernobyl in the Soviet Ukraine. Fission power suggested concrete containment buildings and millirems, high electric bills and half-completed power plants, evacuation plans, radioactive waste, and meltdowns. The fusion community could not afford to be linked to all of that. Fusion's image had to remain pure. No accidents, waste problems and cost overruns were among fusion's core commandments.

"We have to be safety conscious," said J. R. Thompson, the former

NASA rocketry administrator who became Furth's deputy at Princeton. "We have to learn from what happened to fission or we will carry their albatross."

Unlike fission, a fusion reaction is benign in that it does not produce any radioactive byproducts, just inert helium and speeding, free neutrons. The radioactive material is in the fuel alone – more precisely, in the addition of tritium to the fuel mix. Tritium has a half-life of twelve years, very short compared to other radioactive fuels. The fission reactor fuel uranium 235 has a half-life of 713 million years. Plutonium has a half-life of 24,360 years.

Tritium is a more efficient fusion fuel than simple hydrogen because it carries two extra neutrons in its nucleus (deuterium, another form of hydrogen, carries one extra). The higher number of neutrons in the fuel mix increases the chances for the fast-moving particles in a superheated plasma to collide and fuse.

While deuterium and hydrogen are found in seawater, tritium is rare. It can easily be manufactured, however, by allowing neutrons to strike the element lithium. Fusion reactor designers have envisioned surrounding the reactor's vacuum vessel with a "blanket" of lithium to serve two ends. It would generate tritium for fuel and capture the heat of errant neutrons speeding off the fusion reaction. The heated lithium would create steam to drive the electricity generators. In all a reactor would have about twenty pounds of tritium on hand, stored in a special double or triple sealed containment system.

Lithium carries its own hazards, however. This highly volatile element explodes on contact with air or water. A safe reactor design would have to carefully shield the lithium blanket from the reactor's cooling system. But the use of solid lithium compounds in the blanket instead of pure lithium and the use of helium coolant could greatly reduce the risk, fusion planners asserted.

Fusion reactors could run without tritium, but researchers do not yet see how to make a less powerful hydrogen or deuterium reactor economically attractive.

Tritium fuel was not the only door through which radioactivity entered the fusion scheme. The reaction, although benign, would spew off millions of neutrons every second. Most of these uncharged particles would pass through the metal walls of the magnet-encased vacuum vessel and into the lithium blanket. But many would strike the vessel walls and, over time, alter the atomic structure of the walls under the bombardment. Indeed, the alteration would leave them mildly radioactive. The vessel walls would be replaced through the use of remote handling devices, robotics, perhaps every five years, researchers estimate. Most of this radioactive waste would have to be buried for perhaps 100 years before it could be handled.

These harsh facts were not hidden by the fusion scientists, but neither were they emphasized.

The issue of fusion's radioactive baggage has been the focus of some research. The British researchers Roger Hancox and Wallace Redpath issued periodic reports on the question for decades, including a report in 1985 on the safety and environmental impact of fusion reactors, making a comparison with fission.[4] They found the chance of a radioactive gas release at a fusion plant inherently lower than for a fission reactor, but they went on to compare the effects of a major accident. A tritium release would be serious but far less odious than the radioactive heavy metal fuels released in a fission disaster, they said. Tritium does not concentrate in the body the way the heavy metals do and disperses much more rapidly through the environment than most fission products. For example, the upper layers of soil would be contaminated for a matter of days in a fusion accident as opposed to years in a fission accident.

After the Chernobyl fission plant disaster – which stopped short of a total meltdown – fallout containing the radioactive fission products iodine and cesium posed the greatest health threat. According to Lynn Anspaugh, an environmental scientist at Livermore lab, the total cesium emission from Chernobyl was about 1,000 times greater than that emitted in a 20-kiloton nuclear blast such as the Hiroshima bomb.[5] Fusion produces neither iodine nor cesium.

Examining normal reactor operation, Hancox and Redpath noted that, although the radioactive intensity of the fusion waste was relatively low, the actual tonnage of waste was still high. The two researchers found that a particular U.S. fusion reactor design called Starfire might produce about 310 tons of radioactive waste per year, most of it from the inner wall and the lithium blanket. "Available studies indicate that the volume of active waste generated by fusion reactors will be comparable with that from fission reactors," they reported. This was followed by an explanation that fusion waste "is much different in composition" from fission waste, containing no pernicious heavy metals or fission products. Moreover, by using proper materials for vessel walls, the waste could be handled safely in less than 100 years, as opposed to the many thousands of years fission waste takes to shed its radioactivity.

The litany of the fusion faith had always contained mainly cheerful messages: endless energy that was environmentally attractive; a safe reactor with no weapons potential; and a better and safer alternative to the burning of coal and oil or the splitting of uranium. But in truth fusion's vaunted attributes were all relative.

Yes, fusion could produce virtually limitless energy, but at a high cost. It was clean energy only in comparison to fission wastes or the atmospheric pollution of burning fossil fuels. Yes, it was safe energy in that no meltdown was possible, but it could be made part of a fusion-fission system. No, it was not a weapons program, but its science could be applied to nuclear weapons design. All of these caveats took some of the burnish off the utopian fusion vision.

Harold Furth was a patient and calm man. But at the end of a Princeton lab tour he conducted with skittish local political leaders, a certain edge came into his voice. "Hard choices will have to be made," he told the group, "between allowing our standard of living to go down or tolerating a certain amount of radioactivity. There aren't any happy, cheerful solutions."

12
Struggling to sell fusion

Until he cried out that fusion was destined for commercial failure, Lawrence Lidsky was not well known in fusion circles. A red-bearded professor of nuclear engineering at MIT, Lidsky had been quietly studying fusion and fission reactor schemes for twenty years. In the early 1980s, Lidsky came to the conclusion that because magnetic fusion was being developed in large, complicated machines it could never compete economically with existing fission power plants. He thought instead that small, modular, nonbreeding fission reactors, burning natural uranium extracted from seawater, could be the ideal long-term solution of the world's energy problem. Lidsky felt his views were not being taken seriously enough and that physicists were unwilling to discuss fusion's economic problems. So he took his criticisms outside the fraternity, to the newspapers. In a field that had never before experienced a dissident, the fusion fraternity was shaken by Lidsky's 1983 article, "The Trouble With Fusion." For a time it seemed that all hell had broken loose.

The piece first appeared in *Technology Review,* a magazine edited at MIT.[1] But a month later, Lidsky's adaptation ran in the *Washington Post,* hometown paper of fusion's financiers in Congress and the U.S. Department of Energy. The *Post* headline was devastating: "Our Energy Ace in the Hole is a Joker: Fusion Won't Fly."[2] When the Associated Press news agency picked up the story, Lidsky's cry was heard all over the country under such headlines as "Atomic Fusion Called Dead End," "Fusion Called Easy and Useless," and "Fusion:

Veteran of 'Ideal' Energy Program Says Problems are Insurmountable."

Lidsky did not mince words. "Producing net power from fusion is a valid scientific goal," he wrote, "but generating electricity commercially is an engineering problem. The requirement is to develop a power source significantly better than those that exist today, and D-T [deuterium and tritium] fusion cannot provide that solution. Even if the fusion program produces a reactor, no one will want it."

No one will want it. That sentence in stark white-on-black letters was paraded across the front cover of *Technology Review* and reverberated in the public debate that followed. Lidsky had not really broken any new ground about the difficulties of building a fusion reactor. Ten years before, in a series of conceptual studies of a tokamak reactor, Gerald Kulcinski and his colleagues at the University of Wisconsin had touched on all the same facts. But with simplicity, forcefulness, and pessimism, Lidsky made his case to the public.

The articles opened a serious debate in the United States about the commercial feasibility and appeal of fusion. Once the science is conquered, how much would a commercial fusion reactor cost? Who would buy it? How would it stack up in the marketplace against other ways of making electricity? After decades of research, the fundamental question for fusion, as Lidsky saw it, was not really "Will it work?" but "Will it sell?" Fusion energy advocates had touted its abundance and safety, but without an eager customer, be it the government or the utility industry, no amount of good will could bring about a reactor.

If Lidsky's complaint seemed a bit premature, there was at least a grain of truth in his implication that fusion could count few economic allies outside government circles. Other than Gulf's General Atomic subsidiary and KMS Fusion Incorporated, a laser research company, there was virtually no private money behind fusion research. Campaigners for the fusion mission often ran into marketplace barriers as formidable as the scientific barriers. Among the few private citizens in the United States who sought to promote fusion, one published a magazine featuring nude women and another was a fringe politician and paranoid who insisted that the Queen of England was an accomplice in the international narcotics trade.

The laser fusion target chamber of Nova at Lawrence Livermore National Laboratory in California. Courtesy LLNL.

Lidsky's motivation for writing the article was to save the national fusion program from a "dangerous path." The inability to deliver what fusion researchers had boldly promised – cheap, inexhaustible electricity – would "disillusion friends and allies," he said, and make it impossible to finance renewed research into different ways of producing fusion. Lidsky felt that the fusion reactor goal should be abandoned for now in favor of a return to basic plasma physics.

"It was a very serious question of what you were promising the public, what you could achieve and when we would get found out," he said later. Confronted with "fundamental" technical problems, the fusion scientists reacted with what he called "the standard physicists' hubris, the feeling that if the physicist puts his mind to anything he can solve it."

Lidsky's main technical points were these: The cost of building a fusion reactor would always be greater than an equally powerful fission reactor because of the fusion reactor's greater size and complexity. Fusion had engineering problems, as well, starting with the radioactive deterioration of the inside wall of the vacuum vessel. Then there was the question of the volatile lithium blanket around the vacuum vessel. The superconducting magnets outside the blanket, cooled to near absolute zero for efficiency, would be sensitive to heat and radiation damage unless they were securely shielded from the hot plasma and blanket. Add to this the magnetic forces wrenching the machine and, Lidsky concluded, "All in all, the engineering will be extremely complex."

Although fusion is inherently safer than fission, Lidsky wrote, fusion's complexity would probably result in small nonnuclear accidents, with shutdowns that could cause financial problems for the operating utilities. He also sounded a grave warning about fusion reactors and weapons. "One of the best ways to produce material for atomic weapons would be to put common, natural uranium or thorium in the blanket of a D–T reactor, where the fusion neutrons would soon transform it to weapons-grade material," he wrote, describing in essence a fusion-fission hybrid reactor put to military use. "Such a reactor would only abet the proliferation of nuclear weapons and could hardly be considered a wise power source to export to unstable governments."

As Lidsky saw it, using deuterium–tritium fuel, with its extra neutrons, was the best way to make abundant fusion but the worst way to make a reactor. "The scientific goal turns out to be an engineering albatross," he wrote. "A chain of undesirable effects ensures that any reactor employing D–T fusion will be a large, complex, expensive, and unreliable source of power." Lidsky's solution? A reactor based on the fusion of protons in the light elements lithium or boron would produce few or no neutrons, but, he conceded, "At present, we do not know how to build a reactor to ignite such fuels. There is no clear path for an alternative scheme, and almost no support."

The reaction to this maverick researcher was swift and substantial. Lidsky's bosses at MIT responded with a scientific rebuttal in *Technology Review*. Later, Lidsky was quietly stripped of his title as an associate director of MIT's Plasma Fusion Center. A more public and protracted riposte, however, came from Princeton's Harold Furth, director of the

United States' flagship fusion lab. Furth shot off a barrage of letters to engineering experts, to MIT's fusion director, to *Technology Review* and to *The Washington Post,* seeking to destroy Lidsky's every argument. He also persisted in an open, eight-month correspondence with Lidsky, attempting to wrest concession on each scientific point.[3] The correspondence was read into the Congressional record during a subcommittee hearing on energy research and production. A set of the letters even made its way to fusion's European branch at JET.

Furth talked about finding improved alloys and structures to make the vessel wall more resistant to neutron damage. He presented arguments to show that reactor efficiency in fission and fusion were roughly comparable and that fusion offered the marketable advantages of minimizing air pollution, safety hazards, and problems of radioactive waste disposal. He stated that engineering problems associated with radiation damage and heat transfer were comparable for the two kinds of reactors. Cheaper operating costs stemming from fusion's low-cost fuel from seawater would help offset capital costs. Furth argued that it was much easier to make weapons materials using existing fission reactors than would ever be the case for the as yet unbuilt fusion reactors. He said that a misuse of fusion reactors would be easy to detect. And he found "unreasonable" the suggestion that the program turn from the known success of deuterium-tritium schemes to the unworkable boron idea.

As the debate raged, Lidsky started to sound rather apocalyptic. He closed a January 13, 1984, letter to Furth with this: "This status quo cannot last. As I'm sure you are aware, all the auguries indicate that the end is near. There isn't much time left to lead the program to scientifically defendable terrain. The further we go in the current direction, the harder the crash will be." Lidsky grew more righteous and Furth grew more arch. During a discussion session on Lidsky's themes at an American Physical Society convention, Furth asked him, "If they found the cure for cancer would you write 'The Trouble With Chemotherapy?'"

It is difficult to know whether the general public took Lidsky's warning to heart, or even understood it. His articles did stir discussion within Congress, the Department of Energy, and the laboratories. And there was little doubt that Lidsky became a pariah in the fusion community. At the conclusion of the controversy, Lidsky was left a wounded and bewildered man. His technical arguments had been strafed by fellow scientists who accused him of carrying a dangerous and communicable disease, pessimism. Didn't the Wright brothers first build an awkward airplane before the sleek technology of jet engines came along? So it would be in the quest for a fusion reactor, the believers explained. The first reactors would be complicated and awkward, but they would lead to better things.

The nation's fusion research program did not immediately change direction in the aftermath of "The Trouble With Fusion." Instead, Lidsky did. For two decades he had had faith in a fusion future. Now he gave up the ideal to work on advanced fission reactors, in his opinion an imperfect but more practical technology. In time his name faded from the headlines, but the echo of his message lingered.

Relatively few individuals outside the scientific and government fraternity worked on their own for a fusion future. Some who did considered it a personal discovery, even a personal crusade. In the federal belt-tightening 1980s, most of those who tried in their own distinct way to promote the research eventually succumbed to the apathy of the marketplace. They were a quirky collection, displaying virtually nothing in common except a desire to advance the cause of fusion energy.

Bob Guccione, the founder and publisher of *Penthouse* magazine, came to fusion in the great entrepreneurial tradition of America, with an eye on a quick profit from an historic technological breakthrough. He had a colorful image, one that he helped perpetuate by wearing his shirts opened to mid-chest to show off ropes of gold chains and medallions. He had a marbled Manhattan townhouse decorated in Renaissance splendor with original paintings by the masters. His net worth was about $250 million, he said, and he was always looking for serious projects in which to pour the steady cash flow from his magazine.

When fusion caught his eye, Guccione became the most committed private investor in the history of the research. Over the course of five years during the 1980s, Guccione spent $16 million to support research into an offbeat miniature tokamak design that promised to revolutionize the world's energy economy. It was a venture capital risk that could have given Guccione more influence than OPEC had it succeeded. His experience illustrated the marketplace hurdles that fusion faced beyond the scientific challenges.

Guccione had more than a passing interest in science. In 1979, he launched a new publication, *Omni,* a magazine of science fact and fiction. After reading an *Omni* interview with a fusion physicist and innovator named Robert Bussard, the wealthy and restless Guccione became personally involved in the fusion mission.

Bussard had worked for Los Alamos lab and then as an assistant to DOE's Bob Hirsch before striking out on his own. He had started a private company incorporated in Maryland, International Nuclear Energy Systems Company (Inesco), through which he hoped to build a small, compact tokamak using extremely powerful electromagnets. Bussard's partner in the design work was the Italian Bruno Coppi, who had found no money to support his little invention in the government program at MIT, his home lab. The compact tokamak would generate tremendous heat in a very small space. The large tokamaks like Princeton's used costly shielding to protect expensive electromagnets from the damaging neutrons of fusion reactions. Bussard and Coppi's innovation sought to do away with the shielding and make the tokamaks small, relatively cheap and disposable. The interior vessel, once it became radioactively weakened by the fusion reaction, could be thrown away and replaced by another modular unit, much as one replaces a burnt out lightbulb.

Guccione recalled thinking that, economically at least, this made absolute sense. He wanted to hear more, so Bussard was invited to dinner at the publisher's New York townhouse.

Over the meal, Bussard described the difficulties he was having obtaining government funding. After all, the Department of Energy had committed $314 million to the tokamak at Princeton and $100 million to the mirror machine at Livermore. Guccione urged Bussard to seek industrial and business investors, but Bussard had already exhausted that route. The trouble was, Bussard told Guccione, he could not stir any enthusiasm because he could not say definitively when a marketable product would be ready. Bussard believed he could produce a commercially viable minireactor in ten years, but there would be no guarantee.

Bussard was unaware that his host was already a true believer in fusion. Guccione said he had long since concluded that fusion was "the only way to go" to solve man's energy problems, and he felt it was absolutely crucial to the future security and technological strength of the United States that it achieve fusion power first. "Who creates the first fusion reactor literally controls the world's energy supply," Guccione said in an interview, "and if it wasn't this country, who was it going to be? Russia? Communist China?" For an individual investor, fusion held a golden promise, Guccione believed. "Imagine having a unique patent on the telephone system and the electric light system combined, because the whole world uses it, especially Third World countries," he said. "It would totally transform the world."

Bob Guccione decided to finance the minitokamak project, hoping that his personal weight might serve as the gravity to draw in other investors. He was not naive about the prospects. The conventional wisdom held that a fusion reactor would take two or three decades to build, not the ten years Bussard was predicting. And the research would cost a few hundred million dollars. But if it worked, the consequences would be nothing short of spectacular.

In March 1980, Guccione formed a partnership with Bussard and turned over, as he recalled later, some $400,000 in startup funds. Engineers, computer programmers, and metalurgists were hired, and Inesco set up a new shop in La Jolla, California, with eighty-five employees. Over the next four years, as design work progressed and the search for investors continued, Guccione poured in $16 million or $17 million, by his accounting. Predictably, the Inesco scientists who attended international meetings endured considerable ribbing about working for one of the most successful purveyors of adult magazines in the world. Physicists and pinups seemed so hilariously incongruous. But the Inesco team knew Guccione was a serious investor and a sincere proponent of fusion. That's what really mattered.

Guccione saw the incongruity, too, and he was not without a sense of humor about it all. To oversee *Penthouse's* interest in Inesco, the publisher created a subsidiary, which he dubbed Penthouse Energy and Technology Systems, thus creating the acronym PETS. It was a conscious reference to *Penthouse's* nude centerfold, "Pet of the Month," and a way for Guccione to acknowledge the uninhibited women who, in truth, were creating the profits to finance fusion research.

Even Guccione and his PETS money, however, could not convince other investors to commit themselves to fusion. Inesco's project was just too speculative. Guccione said he also grew to suspect that the nuclear power industry, the fission plant owners, were putting pressure on Washington to ignore projects like Inesco's, which threatened to replace fission power. Guccione had no direct evidence of this, but it was a theme commonly sounded by frustrated fusion researchers.

In 1984, an attempt to take Inesco public flopped after the underwriter failed to sell the last 400,000 shares. Bussard's dream and Guccione's gamble were crushed. The two men salvaged a few patents that might someday prove valuable and took with them an undying hope for fusion's future. That hope lay with government, Guccione had to concede. Only national governments possessed the resources and the freedom to invest in research projects at such a basic stage and with such expensive tools.

If only the government would mount a "flatout, do-or-die, Manhattan-style project that we had for the development of the nuclear bomb," Guccione lamented. With that kind of commitment fusion could be conquered in a decade or even less, he believed. Fusion would be "the single biggest boon to mankind ever," he said, "the sort of thing a president should get behind." Like the space program, fusion could "put this country right back on top. . . . the number one industrial nation in the world."

Two years after the Inesco collapse, Guccione had not lost his fervor. Fusion is "the ultimate source of energy for this plant," he said. "It obviously can be done. If it exists in nature, it can be recreated by man."

He spoke with as much conviction as the fusion pioneers. Guccione had the litany down pat. Fusion was clean. Fusion was safe. Fusion was a natural process. Fusion was inevitable. He had become one of them, one of the fusion proselytizers, a believer in the mission. Only the money fell short. Outside of the U.S. Department of Energy, there was simply no market for fusion.

If the economics of fusion were in question, perhaps its environmental advantages might be worth something to a pollution-threatened society.

That was the perspective of Ansel Adams, the legendary landscape photographer who, in the last year of his life, became an ardent advocate for fusion.

For thirty-seven years Adams had served on the board of the Sierra Club and had dedicated much of his energies to trying to preserve in painstakingly printed black-and-white photographs the grand vistas of America that he encountered. Although he opposed nuclear weapons, Adams favored nuclear power as a way to create electricity with the least damage to nature. He saw fusion as an environmental imperative, a way to save the landscape from the pollution of conventional energy plants burning coal and oil.

In May 1983, *Playboy* magazine published a lengthy interview with Adams in which he described the views he held on nuclear power that had caused him to break with traditional environmentalists. Of his simultaneous support for conservationism and nuclear power, he told *Playboy:*

> That's an apparent dichotomy and it disturbs a lot of people, but the danger of nuclear power is conjectural and the pollution potential, compared with the known pollution potential of burning coal and oil, is minute. When you consider the threat of acid rain and the general pollution of air and water caused by thermal power production, it is terrible. There is general agreement that nuclear weapons are absurd, but I disagree with the view that nuclear power is bad.

Then Adams put in a plug for fusion:

> The better alternative to the fission reactors is fusion, which the government isn't pushing the way it should. It is a much safer alternative. It's clean, efficient and not very expensive. The technology is inevitable. It's a necessity if we are going to avert a disaster. I just can't be scared. Everything is a risk. When there is a big public squawk about fusion, it becomes evil. It is unfortunate that it has been clumped together with something as insidious as nuclear weapons, because utilizing nuclear energy is the future.

Officials at the Lawrence Livermore National Laboratory in California were surprised and delighted by the Adams interview. Ken Fowler, Livermore's magnetic fusion director, promptly invited Adams to tour the lab's fusion devices. Adams made two visits to Livermore that year, first to look and then to photograph. According to Fowler, the second visit was spurred by Lawrence Lidsky's controversial article, "The Trouble With Fusion."

"Ansel called all upset about the article," Fowler recalled. "He viewed all such things as just negativism and asked what 'we' could do about it. I said, 'Well, you take pictures, don't you?'" Adams complied.

One of Adams' efforts to sell fusion was directed at the president of the United States. The photographer had been vituperative in criticizing his fellow Californian, Ronald Reagan, and the administration's environmental policies. After the *Playboy* interview, Adams was invited to meet with the president. The highly publicized meeting took place on June 30, 1983, at the Beverly-Wilshire Hotel in Los Angeles, and it ranged over many environmental issues. Adams made one of his sharpest thrusts in favor of fusion energy. He wrote in his autobiography,[4]

> During the visit with Reagan, I suggested he take $10 billion from his defense program and apply it to a crash program for magnetic fusion development. Reagan raised an eyebrow at my temerity, but I believe it is obvious that once fusion power is achieved, the energy shortage will be past and we will be independent of foreign fuels. In 1902, the automobile was in its infancy and the airplane an insubstantial dream. From the two-cylinder gas buggy to magnetic fusion is a giant stride, but incredibly it can be accomplished during one lifetime.

Unfortunately, Adams' lifetime was not enough. He died in 1984. In his last interview, with *San Francisco Focus* magazine, the eighty-two-year-old conservationist again pushed for the development of fusion energy. "The point I want to make is that fusion development isn't that far down the road," he said. "It's only as far as our leaders want to make it."[5]

Ansel Adams had been the best spokesman the fusion fraternity could have asked for, but he discovered fusion too late in life to have a major impact. The photographs he took of the Livermore fusion machines disappeared into the Adams archives. But by paying attention to the fusion scientists' cause – even just for a year – he had given them heart. His endorsement as a man of stature outside the scientific community seemed in some small way to validate their work. At Livermore, the grateful fusion missionaries mounted portraits of this cherubic old man with a bristly white beard, lariat tie, and broad-brimmed Stetson where they could see his face every day.

The physicists were delighted to proclaim the support of a giant like Ansel Adams, but they did not know quite what to make of it when they found themselves in the embrace of Lyndon H. LaRouche, Jr.

LaRouche was an ideologically complex character who had a conspiratorial, apocalyptic view of the world. His philosophy had evolved through the years from left-wing Marxism all the way over to archconservatism. Paranoid rhetoric was a mainstay. His criticisms of United States government policy were frequently focused on individuals in a very personal way. Henry A. Kissinger, the former national security adviser and secretary of state in the Nixon Administration, was a favorite object of LaRouche's invective. And, in a 1984 bid for the presidency, LaRouche spent half an hour in a nationally televised advertisement haranguing candidate Walter F. Mondale and accusing the former vice-president of being an "agent of influence" for the Soviet Union.

What's more, Lyndon LaRouche, Jr. believed in fusion.

In 1974, his interest in the technology had prompted him to found the Fusion Energy Foundation. In the early 1980s, whenever candidates spawned by his organization, the National Caucus of Labor Committees, ran for local offices they were likely to mention fusion as a key plank in the platform, an energy source that could build a secure America. Using volunteers from among the LaRouche faithful, the Fusion Energy Foundation solicited funds at airport terminals around the country and continued to spread the word about fusion. Management at several airports tried to throw out the LaRouchers, but a string of court decisions preserved the foundation's First Amendment right of free speech.

The leafletting was perhaps the most visible and consistent public relations effort on behalf of fusion energy that the cause had ever seen. But the benefits to fusion's image were questionable. It was hard to know whether the LaRouche brand of salesmanship opened doors or bolted them shut. On February 11, 1982, that point came into sharp focus. Nancy Kissinger and Ellen Kaplan, a Fusion Energy Foundation worker, scuffled at Newark International Airport. According to court testimony and newspaper accounts, Kaplan walked up to Henry Kissinger as the former secretary of state and his wife, Nancy, were about to board a plane for Boston, where he was scheduled to undergo heart surgery. Kaplan asked a question that the Kissingers took to be offensive, and Mrs. Kissinger grabbed her.

In copius accounts, newspaper, television, and radio reporters across the country identified Kaplan as a volunteer for the Fusion Energy Foundation and an advocate of nuclear power. The incident was described again and again over the ensuing months as Kaplan pressed assault charges. A Newark judge eventually acquitted Mrs. Kissinger, declaring that she had displayed a "somewhat human reaction to an offensive question."[6]

A fringe political group, the secretary of state, an altercation, a fusion volunteer – it was not the kind of image building that the fusion scientists hungered for.

Still, the Fusion Energy Foundation served other, more constructive purposes. In testimony before Congress, it frequently argued the merits of fusion research, and its magazine, *Fusion,* at first espoused the cause in a staidly scientific way. Later, it started advocating some weapons use of fusion technology, and LaRouchian paranoia crept into the magazine. The fusion community cringed. When the foundation presented an award to Mel Gottlieb, Harold Furth's predecessor as head of the Princeton Plasma Physics Lab, colleagues elsewhere questioned the wisdom of accepting it. Gottlieb later regretted any association with the foundation and said he felt "used."

In 1987, *Fusion* magazine was ordered to close by the federal government. U.S. marshalls seized its office and bank accounts in connection with a federal indictment for credit card fraud brought against various organizations operated by LaRouche. In 1988, LaRouche was convicted of trying to defraud federal tax collectors by hiding his income and failing to repay more than $30 million in loans from his supporters. He was later sentenced to fifteen years in prison.

Thus did fusion lose its most visible private supporter.

Luella Slaner, the philanthropist from Scarsdale, New York, was touched by the fusion bug in 1973, the year of the first Arab oil embargo. By 1975, she had become a one-woman road show, plugging fusion to anyone who would listen. She, too, had become one of fusion's missionaries. Slaner and her husband, Alfred, grew concerned about America's economic security during the anxious months of the oil crisis. "Without energy," said Luella Slaner, "we can't produce anything, and the United States will just become second rate, third rate in everything. The reason we had survived in the past was we'd had an abundance of cheap energy."

The Slaners' patriotic concern was not passive. First they made small sacrifices. Like many Americans, Alfred Slaner started using public transportation instead of his car for the commute to his job in New York City. He was a relatively wealthy man, executive vice-president of the textiles giant Kayser Roth Corporation. Luella Slaner lowered the thermostat in their home so much that dinner guests were forced to bring sweaters.

When Alfred Slaner learned about fusion from a company scientist, his interest was piqued and he dispatched his wife, the daughter of a Columbia University chemistry professor, to find out more about it. What she learned, she said, was that "plasma physics was a very low-key science. It was not very well funded. It had no sex appeal whatever."

She talked directly to plasma physicists and reported to her husband that fusion was worth supporting. The year was 1975. With a bit of seed money from a small family foundation set up to finance charitable interests, Luella Slaner created the Society for the Advancement of Fusion Energy (SAFE).

"I'm an activist, I believe in getting things accomplished," she said.

Over the years, Luella Slaner wrote thousands of letters soliciting funds, hammering Congress and extolling fusion energy. She fed articles to newspapers and sent speakers around the country. She helped produce an educational film on fusion featuring Neil Armstrong, the moon-walking astronaut. In all, the society spent close to $1 million to publicize fusion – before it fizzled in frustrating failure.

SAFE suspended operations after the oil scare died out and the federal budget blossomed with new military spending in the 1980s. Luella Slaner found that it was almost useless to try to persuade elected officials to support fusion. "No congressman who is in office for two years is going to vote for something that's fifteen years away," she said, "not when he has his constituents clamoring for projects back home. And we have to compete with the weapons."

She concluded that "there's just no public interest at all in fusion, only a very small group of people who know what's being done at the national labs. Rightly so, I guess. They have all the oil and gas they want and no one thinks of the future. I got involved in this not for myself but for my grandchildren. The American people have had such an abundance that they never really felt it was important to plan for next year."

Bob Guccione, Ansel Adams, Lyndon LaRouche, Luella Slaner: Fusion made strange bedfellows, but it did not make many of them. The risk, the cost, and, above all, the time needed to perfect the arcane science were on such a scale that individuals could not hope to propel the mission to success.

Fusion needed friends, advocates in high places, money, and vision. Fusion needed people who knew what it took to look into the future and plan today to meet tomorrow's challenges.

In 1981, one of fusion's best friends in Congress, Democratic Representative Mike McCormack of Washington, explained the fraternity's predicament bluntly in a speech to the Atomic Industrial Forum. He was sponsor of the Magnetic Fusion Engineering Act of 1980, a congressional endorsement of the fusion mission. During the speech, he addressed the 1981 federal budget cuts that threatened to curtail some fusion work:

> Unfortunately, such a tragedy occurs because the fusion program has almost no constituency. Most of what [constituency] it has is dependent upon DOE funding and therefore reluctant to speak out. I think the time has come for a public outcry for this unspeakable folly and this thoughtless budget cutting for a program that will provide energy self-sufficiency for this country early in the twenty-first century, if not before. In a very real way, fusion is everybody's problem, and everybody should be concerned.

In the 1950s the nation's utilities had taken an early interest in fusion when a powerplant was said to be just around the corner. A group of Texas utilities, for example, spent close to $1 million a year on fusion research at General Atomic and then, in 1956, shifted that support to the University of Texas. During the energy crisis of the 1970s, some token amount of utility money again found its way to the fusion labs. The Electric Power Research Institute (EPRI), an industry group, started a fusion program to help insure that the fusion reactors being designed would speak to the utilities' needs.

During the 1970s, EPRI spent as much as the federal government on studies of fusion-fission hybrids, according to F. Robert Scott, a former General Atomic researcher who became fusion program manager when EPRI was formed in 1974. The utilities saw the hybrid breeder as a way of keeping down the price of scarce uranium. The energy crisis also spurred Public Service Electric & Gas Company, the New Jersey utility that supplies the Princeton lab with its power (at an annual bill of $6 million), to begin grants of about $50,000 a year to the lab and to carry out some in-house fusion research. Texas Utilities continued donations to university level fusion research and the New York area utilities contributed to the University of Rochester's unclassified laser fusion work.

But, in the 1980s, an industrywide retrenchment by the power utilities caused a turnabout in their commitment to energy research. "The Institute had its budget crunch," said Scott. "Fusion was too far out in the future. There was not much EPRI could do by putting $1 or $2 million into the program. The utility industry changed its idea of what its responsibilities are for research. It was a much more short-term viewpoint." In 1985, the Electric Power Research Institute eliminated its fusion program.

In 1980, Phillips Petroleum Company, looking for alternative energy schemes as a hedge against depleting oil supplies, made a tentative foray into fusion. Gulf Oil Company, owner of General Atomic in San Diego, was looking for a partner to help subsidize work on a compact, circular pinch machine designed by Tihiro Ohkawa. He named it the ohmically heated toroidal experiment or OHTE and pronounced it "oh-tay" meaning "checkmate" in Japanese. It intrigued the Phillips research and development division as potentially better suited to commercial uses than the large, complex tokamaks. Phillips became an equal partner in OHTE and contributed approximately $15 million to get experiments started on the project.

But just as Phillips and Gulf were about to commit themselves to a

$150 million upgraded version of OHTE, the two oil companies suffered systemic economic shakeups. Phillips crippled itself with debt fighting off a takeover bid by T. Boone Pickens and chose to jettison the speculative and long-term fusion research. Shortly thereafter, Gulf was taken over by Standard Oil of California (Chevron) in another deal that had Pickens' stamp. Standard Oil promptly sold General Atomic, now called GA Technologies, to a private investor named Neil Blue, and the OHTE upgrade was never built.

Industry was not yet willing to commit itself to fusion. Who in America, aside from the plasma physicists, wanted fusion energy? It remained a utopian technological goal that for the present required a leap of faith. For dispassionate investors with a bottom line, fusion still appeared too risky.

Only national governments stood a chance of making fusion energy a reality. But in the 1980s, astounding deficits in the U.S. federal budget and an accompanying reexamination of national research policy sent the fusion community in the United States into a fiscal and psychological tailspin.

Certain realities had to be faced. Fusion as an energy source for the twentieth century was now obviously out of the question. The twenty-year outlook for success described in the 1950s, then again in the 1960s and 1970s, had become a joke, even among the scientists. Fusion was a very reliable science, they had said. A reactor was always just twenty years away.

With declines in oil and uranium prices, the crisis atmosphere for fuels lifted, and the sense of urgency about fusion and other alternative energy schemes faded. In the final year of the Carter Administration, a last gasp of energy policy initiative had produced the Magnetic Fusion Energy Engineering Act of 1980. It authorized a $20 billion drive to build a commercial-sized demonstration fusion reactor by the year 2000. But the act proved to be little more than a declaration of intention. Congress and the Reagan Administration, citing budget constraints, quickly scuttled the plan and, as part of general trimming of research and development programs, began to cut back on fusion funding.

At the same time, the president's Office of Science and Technology Policy began asking new questions. Should government really be in the business of building prototype nuclear reactors? Was that not the job of the private sector? Funding basic research, as opposed to applied science, was a more natural role for government.

The Fusion Engineering Act had envisioned three steps to commercial fusion, all involving a major federal role. First would be the proof of scientific feasibility, the breakeven experiment at Princeton. Then a new machine would be built, the Fusion Engineering Device, to demonstrate the engineering feasibility of fusion energy. Finally, a demonstration powerplant would be built. At that point, presumably the utilities would take over.

In the new budget atmosphere, however, the DOE intended to limit government involvement, stepping out after building an engineering test reactor. Industry should then have enough information to assess the potential for commercializing fusion energy, it was argued.

In June 1984, George Keyworth, the president's science advisor, stated the new energy research policy of the Reagan Administration quite succinctly, "The primary role of the federal government in science and technology is to support high-quality basic research," he wrote in the technical journal *Research and Development*. "Except in such areas as defense and space, where the government itself is the user of the technology, development can be performed more effectively by the private sector."

The fusion researchers were vexed by the new signals. Now that they were within a stone's throw of breakeven, their first major scientific goal, the financiers were quoting the rules of the marketplace. "We were criticized for being starry-eyed visionaries who couldn't do it," said Harold Furth, remembering fusion's long history. "We thought one day the world would beat a path to our door. Instead, they've just stopped saying we couldn't do it. Now they say, who wants it?"

Beginning with the Hirsch era, the fusion program had been pushed to show engineering progress in a series of ever bigger research machines. Now Keyworth was suddenly pressing to slow down the pace of fusion machine construction in favor of a better understanding of the physics.

"Everything that used to be white is now black. Everything that used to be black is now white," Furth lamented in the critical summer of 1984. "Now they believe with a vengeance everything we used to tell them."

Furth and other fusion scientists, however, did not see the change of emphasis as a prejudice against fusion per say, but rather a general result of difficult economic times that undermined all energy research and development programs. Solar, synfuels, breeders, and what Keyworth called "the trendy energy research projects of the 1970s" had long since been struck down. Of all of President Carter's alternative energy programs, only fusion was left.

Into this highly unstable environment entered another grand and expensive scientific goal, that of smashing an atom into all its constituent parts to learn more about the universe. *Particle physics,* as it was called, was often confused with *plasma physics* by laymen, and plasma physicists resented it. They always regarded the particle scientists as impractical, artists really, indulging themselves in a treasure hunt for new particles they could then list and win Nobel Prizes for.

The particle physicists were well organized and knew how to market their indulgences. They often appeared in the newspapers, usually on the front page. Of course, in their search for subatomic particles, they too were asking for more money to smash more atoms in underground accelerators. The quark, the lepton, the neutrino – the names they gave to their new-found particles – were fascinating, no doubt. But for what purpose? the fusion scientists asked. Still, they had captured the imagination of the American public as fusion never had.

One afternoon in January 1986, three years after the startup of the new Princeton tokamak, James Cronin, Nobel Prize winner and one of the better known particle physicists, came to the Princeton Plasma Physics Lab to give a lecture about his field and the new machine his colleagues wanted to build – the superconducting supercollider known as the SSC.

The lab's auditorium was filled with several hundred curious and skeptical plasma physicists. Cronin's speech was not so much technical as inspirational as he tried to address the question "Why build the superconducting supercollider?" and "Why study particle physics?"

Cronin quoted Pascal who, in his *Pensees,* said that "all he [mankind] can conceive of nature's immensity is in the womb of an atom." Cronin spoke further of a search for a theory that would unify the laws of electricity, magnetism, and gravity, a theory that might be revealed by breaking matter down into its most minuscule parts. At Fermilab outside Chicago where he worked, speeding particles were smashed head on in a circular tunnel four miles in circumference, the particle accelerator. Now the particle physicists were pushing the president and Congress to approve the largest particle accelerator ever, even bigger than the one at CERN, the European Laboratory for Particle Physics near Geneva, that was being expanded to stretch for sixteen miles underground.

The American scientists were proposing a fifty-two-mile oval accelerator, big enough to encircle the city of Washington, D.C. as Cronin showed on a projected map.

Cronin had been speaking for about an hour on quarks and leptons and the big accelerator, when he paused to take questions from the plasma physicists. A hand went up in the audience and a dry voice called out, "How much is this accelerator going to cost?"

Cronin replied that, according to a May 1984 study, costs would total $3 billion, maybe a little more. Laughter rippled across the auditorium. It was the laughter of black humor. The entire budget for the Princeton lab was about $100 million a year. The most expensive fusion machine ever built in the country, the one sitting in their basement, had an official cost of merely $314 million and Washington had begrudged them every penny. The Princeton physicists laughed bleakly because they knew there was a good chance that Dr. Cronin would get his $3 billion. (Later official projections put the cost at $8 billion.)

Cronin responded to the laughter. "Aside from its rather large cost it seems very feasible to build," he said to more tittering. "Bigger is better because it's the way we can make progress. It's a question of what you want to do about science," he pleaded. "It will not benefit society in a material way in the near future. If you're concerned about a visible return to society, we have a very weak case. But if the SSC is not built here the focus will shift to Europe, and American physicists will move there. A great nation like this has no excuse for not doing frontier science in all fields. In the long tide of history more knowledge has been helpful."

To Don Grove it seemed that the accelerator scientists succeeded by flaunting the impracticality of their research. It was the practical aim of fusion research that at times dragged it down. When the physicists paused to do intricate experiments on their machines, they were accused of "playing in the sandbox," not moving quickly enough toward their ultimate goal of a reactor. When they pushed ahead to build new machines without a strong theoretical basis, they were accused of gambling with the government's money. They couldn't win. Commercialization dogged their every decision.

Erik Storm, the Livermore laser fusion leader, shared Grove's frustration. "I think it is more important for the future of this planet to study fusion energy than to study the color of quarks," he fumed.

Oh to be among the particle physicists, left alone with their billion dollar machines to contemplate the nature of matter and energy, to search for truth and beauty. It seemed like no one was ever going to win a Nobel prize for fusion research. It was applied physics, too practical to be considered by the judges. You don't win a Nobel prize for building a bridge.

In the summer of 1984, Congress chopped the Reagan Administration's $483 million 1985 budget request for fusion down to $440 million. It was the first downturn in fusion funding in the field's history. The DOE and the lab directors were faced with postponing the big experiments or snuffing out many of the smaller support projects. So Princeton's long-awaited breakeven test, scheduled for 1986, was ordered put on hold for another two or three years.

But Princeton's woes were petty compared with the tough news the DOE delivered to the mirror machine researchers at Livermore the following summer, when the fusion program endured another cutback. The huge Mirror Fusion Test Facility, which had just been completed at a cost of $372 million, was to be mothballed. Not one experiment would be done.

The verdict was delivered to Livermore just a few weeks before the giant mirror's formal dedication was scheduled. Contingents from all the U.S. labs were expected for the ceremony that marked the end of eight years of construction. When Livermore announced that the dedication would go on as planned, guests felt as though they were heading to the West Coast for a funeral. People from Princeton, where the big tokamak's survival had meant the mirror program's demise, were especially uncomfortable.

"I thought I was going to a wake," said Mary Shoaf, Harold Furth's administrative assistant. "Harold and I really dreaded going out there." The tour of the comatose mirror machine was "kind of eerie," she said, "walking under the device, the water dripping down from the refrigeration units. There were very few people there. It was a monstrous device with all the safety nets strung along the sides like a great ship stranded in its rigging."

Ken Fowler, Livermore's magnetic fusion director, and the nominally retired Dick Post were determined to salvage some role for the mirror program. Eventually, Fowler helped persuade the DOE to spare one small mirror device at the University of Wisconsin called Phaedrus and MIT's modest mirror machine. But the overall devastation of the mirror program before it had a chance to reach for breakeven sapped Post's spirit. The man with the crinkly eyed smile spoke with an unaccustomed bitterness about the turn of events.

The mirror had not been given enough time to mature, Post argued. Tokamaks had had that luxury. "It took years for Artsimovich to get to where he was prepared to go out and be an advocate (for tokamaks)," Post said. "There's a gestation period for the understanding of these things." As for himself, Post said, "I still have some mirror work to do either theoretically or experimentally and I'll continue to do that even if there's no one around."

The bitterness in the fraternity was palpable. Even past alliances seemed to be crumbling. In a 1985 speech to the American Nuclear Society, Bob Hirsch shocked and infuriated the fusion community by denouncing the tokamak as an impractical reactor design.[7]

As the former "young man in a hurry," it was Hirsch himself who had launched the era of the big tokamak in the 1970s before leaving government to become a deputy manager of the science and technology department at Exxon and then a research vice-president at Arco. Through that time he had remained a key figure on the DOE's Energy Research Advisory Board. Now, he was saying, years of "heroic efforts" had not removed "the fatal flaw of the tokamak concept: It is inherently a complex maze of rings and a toroidal chamber inside of other rings. In my view, this complex geometry will not be acceptable to the utility world where power plants must be maintained and serviced rapidly and at low cost. In that world, simple geometries are essential."

Hirsch made many of the same observations as Lawrence Lidsky about fusion's engineering difficulties and therefore its commercial problems. Unlike Lidsky, however, Hirsch did not call for a retreat into basic plasma physics. Instead, Hirsch had more shocking advice: Go ahead immediately with a tritium-fueled breakeven experiment at Princeton and then abandon the tokamak and try something else. With the money saved, other, more practical designs should be pursued – designs that would make attractive reactors – even if the current physics knowledge about them was lacking. He suggested increased research into modern "pinch" machines and the Spheromak, a compact torus. He agreed with the mothballing of the giant mirror, suggesting mirror research be pursued at a smaller scale until it could justify such a big leap.

Hirsch's denunciation of the tokamak – the machine he had championed a decade earlier – had worldwide reverberations. He was not an obscure engineering professor like Lidsky but the former head of the U.S. fusion program and currently a member of the president's energy advisory board. In Japan, members of the Japan Atomic Energy Research Institute's fusion planning team were commanded to write an internal rebuttal to Hirsch's speech. Japan, after all, was about to begin experiments on its own giant tokamak and was planning a multibillion dollar successor machine. Hans-Otto Wuster, the JET director, was in the audience for the Hirsch speech and called it "scandalous."

"He is a man who thinks [fusion] has to be in this century," said Wuster. "How does he teach nature that?"

One researcher at Princeton had a typical reaction:

> Hirsch took the science out of the program to begin with. Scientists cringed in the 1970s when Hirsch promised a viable reactor by the year

2000 with his Madison Avenue technicolor viewgraphs. Now he has the gall to say '*You* haven't delivered on what *you* promised.' It's what *he* promised. He says put tritium in TFTR. Get it over with. It's really infuriating.

Hirsch said later that his purpose in making the anti-tokamak speech was to shake people up:

If you take the rose-colored glasses off, all tokamak reactor designs "look like a jumble, a mechanical nightmare. No way on God's green earth will practical people accept anything like what these guys have designed. They just won't do it. Because in the real world . . . things break, they need fixing during operations, particularly high-tech things. There is no way that you're going to be able to cheaply and quickly fix a broken tokamak.

He continued:

The major point I was trying to make, is that people have crowded around the tokamak at almost the exclusion of other ideas and approaches. The problem is that the point of fusion is not to make tokamaks work, but to make a product that is going to be useful, economic, reasonable, and desirable when it's done. It has to be something the marketplace is going to want to have and utilizes. If it isn't, if it just produces something that works, that's not enough. You're trying to make a practical result. I think the program has lost sight of that."

Hirsch was especially irked by the fact that TFTR, more than three years after its first plasma, was not yet running on tritium and reaching for breakeven. "That great big machine in Princeton costs an enormous amount of money to run as a hydrogen machine for hydrogen physics," he said. "It's hard to justify. To do that for three or four years is insane. It means that a whole lot of universities and new ideas are not getting the money they need." Hirsch blamed Harold Furth as much as the DOE for the foot dragging by the Princeton researchers. "Ever since we built that machine they've wanted to do hydrogen plasma physics, period, and at the end they'll do some deuterium and tritium. That was a management issue for me from the outset." It was obviously galling to Hirsch that his successors had not forced Furth and company to do the radioactive experiments for which TFTR was designed but instead had postponed breakeven for budget reasons. "Harold Furth is delighted to be able to do his hydrogen plasma physics in a big machine because that's his world," Hirsch spat.

Like Lidsky, Hirsch said he was only trying to help a program about which he cared deeply. "If you speak up too loudly, you can hurt the program, and I don't want to do that. And I'm not running the program anymore and maybe I couldn't lead the program well today. Time passes all of us by at some point."

Hirsch and Lidsky, Furth and Post, even Bob Guccione, they all wanted fusion to succeed, but they had different ideas about how to get there. Engineering physics or basic physics, tokamaks or mirrors, small machines or big, there were arguments for all these paths to a product that the public and the utilities would embrace. But until that product was ready, the selling of fusion would be no easy matter.

13
In sight of breakeven

A chauffeur-driven car with white doilies on the headrests carried Dr. Yasuhiko Iso out into the flat countryside near Japan's big fusion center in Naka-machi, north of Tokyo. Iso's destination was the local Shinto shrine, where it was his intention to thank the resident god for the recent success of the JT-60 tokamak. A few weeks before, on April 8, 1985, Japan's entry in the breakeven race had fired off its "first plasma" before an audience of newsmen. Then, for the cameras, the scientists painted in the second black eye of a large Daruma doll like the one Iso had given Princeton.

In the seven years of JT-60's construction, the plasma physicists at Japan's Atomic Energy Research Institute had regularly visited the local shrine. And in the factories, as each piece of the tokamak was completed, workers gathered around as a Shinto priest blessed the hardware. In Japan, science and spirituality were not incompatible.

"When we feel we cannot do any more," Iso explained, "when we have done the best we can and yet we are still in doubt, we want to rely on something absolute, and this must be God. Therefore, we go to the shrine."

Iso was the director of the fusion research center – equivalent in rank to Harold Furth at Princeton or Hans-Otto Wuster at JET. He was a slim man in his early fifties with a small, soft face and high eyebrows that gave him a surprised look. He had worshiped at the shrine before the first plasma attempt and felt an obligation now to return.

The country air that day was damp, cool and fragrant with the oils of cedar and pine. The sky was overcast, and a breeze caught the tree branches. After a fifteen-minute drive Iso's chauffeur parked the car at the bottom of a steep bank of stone stairs. Iso labored up 131 steps. Ahead stood the wooden enclosure of the shrine, clouded by pink cherry blossoms blowing down in the quickening wind. They looked to him like snow.

The local deity was the goddess of weaving. At her shrine, Iso tossed a coin into an offering box and then bowed twice toward a circular mirror suspended in the pagoda. He clapped his hands loudly twice, and then bowed again. He examined a rack of flat "wish" sticks on a table, each one lettered differently in red. The sticks indicated wishes for recovery from sickness, happiness at home, success at business or school, general luck, or a special wish. On his last visit Iso had chosen the special wish stick. Today he chose the luck stick. Thunder sounded in the distance.

With a felt pen Iso wrote his name and address on the stick (so that the god would know exactly who was sending the wish) and, on its back, a quick note of thanks in bold ideograms for JT-60's good fortune. The machine's name was written in roman letters. He then placed the stick in a box and bowed once more. He smiled, satisfied, and then contemplated the pleasantly cool, tranquil scene. Towering above were the mammoth, century-old cedars with their thick splintery trunks. Iso paused at a likeness in bluish stone of the weaving woman, then he strode down the stone steps to his driver and reentered the limousine.

As rain came down in splatters, then hailstones, Iso said to a visiting journalist, "I have always been known as lucky – I, JT-60, and the Japan Atomic Energy Research Institute. You are looking at a man of good fortune. You do as much as you can, the best you can, and then there is something else that oversees it all, that takes over." The clatter of the hailstones faded and up ahead the sky over the fusion research center had cleared.

In the years before the giant tokamaks hit their stride, doubts about the machine and about fusion's future had reached another crescendo, part of the cyclical ebb and flow of enthusiasm that is so typical of fusion's history. But as experiments proceeded in the late 1980s, signs of good fortune began to visit the laboratories. As with the Daruma doll, the researchers had faithfully and patiently stuck to their task and were beginning to be rewarded with insight. After so many decades of disappointment, it finally seemed as though the Gods were with fusion.

The years 1985 and 1986 marked the start of another cycle of hope for the utopian science, a hope that extended briefly from the scientific realm to the political.

On the technical front, a handful of tokamaks, including Europe's JET and Princeton's TFTR, suddenly burst out of the limits of the "low mode," the depressed plasma state in which energy simply fizzled. The new, hotter "high mode" was surely the road to breakeven. Another research breakthrough promised meaningful fusion applications. Scientists at A.T.&T. Bell Laboratories and elsewhere in the United States invented a revolutionary ceramic material that loses no energy as it conducts electricity, making it possible to build strong electromagnets that do not waste power or require as much expensive coolant. Smaller, more powerful and much less costly magnets might surround future tokamak vessels, making the tokamak more attractive as a commercial reactor.

The Department of Energy was still waiting for proof of the scientific feasibility of fusion – breakeven and ignition – before committing the government to a prototype commercial reactor. And, in 1986, American fusion teams sought to advance their cause by producing hotter and hotter plasmas – the highest temperatures ever artificially produced in a laboratory. A terse prescription for political success here was offered by Don Grove: "To be able to stand up before Congress and say we've set a world record temperature would be better than anything." Grove knew that although the record temperatures might not be truly meaningful scientific breakthroughs, Congress seemed to get wide-eyed over "bests" and "firsts."

The Livermore lab's laser researchers weighed in first. Before a Congressional energy committee they said they had essentially pulled even with magnetic fusion. Their Nova laser had produced a dense fusion reaction at more than 100 million degrees centigrade, spawning

20 trillion neutrons in 25 trillionths of a second. No controlled fusion device had ever produced so many neutrons so quickly.

The laser fusion people acknowledged that, for certain technical reasons, their brand of fusion would require a much hotter and denser reaction than magnetic fusion. But even with these caveats, their news seemed impressive. Secret results from the Nevada Test Site further emboldened Livermore's Erik Storm to tell Congress that if only the United States would launch "an Apollo-like program where you go all out," laser fusion could have an engineering test reactor by the year 2010 and a prototype power plant ten years later.

In the summer of 1986, good fortune and surprising events at Princeton delivered even more significant results. Since 1980, the temperature record for a magnetic fusion plasma had stood at 80 million degrees centigrade, the achievement of the Princeton Large Torus. Practical magnetic fusion required a temperature of at least 100 million degrees and ideally 200 million degrees at the reactor core – more than ten times hotter than the center of the sun – to compensate for the sun's gravity. Moreover, this energy had to be stored in the plasma for a significant amount of time.

Ironically, the most successful method for heating large tokamak plasmas – the use of neutral beams – caused a relative reduction in confinement time as the plasma got hotter. Princeton's giant tokamak, TFTR, had been stuck in this predicament, aptly dubbed the "low mode," since it was first fired up. TFTR was in danger of becoming a very expensive white elephant. Additional neutral beams, diagnostic devices, and operating costs had raised the price of the device to almost $1 billion, according to Dale Meade, deputy head of TFTR.

All that changed with the unexpected arrival of "supershots." It all began quite innocently when two Princeton physicists decided to give TFTR a good "bake-out" cleaning in preparation for a marginal experiment that had been put off for a year. Jim Strachan and Mike Zarnstorff were given two days to run their show. The pair made an odd couple. Strachan, forty years of age, was a moody and dour man, insistent about his opinions and cynical about the grand promises made for fusion. Zarnstorff – Mike Z. to his colleagues – was a relentlessly cheerful fellow of thirty-three who wore bright short-sleeved polo shirts and bare-toed sandals, even in midwinter.

First, several helium plasmas were produced to burn the impurities off the tokamak's walls – the self-cleaning oven approach. A vacuum pump sucked out the gaseous residue. To reach a superclean state, Strachan and Mike Z. ran a few more shots than usual. They also had the benefit of a new alignment for the neutral beams that, for the first time, allowed particles to be injected in opposing directions, with and against the flow of the plasma.

On the second day the experiment began with hydrogen plasmas of low density heated by neutral beams. After a few shots, however, the physicists manning the diagnostic devices began to notice something extremely peculiar. The electron temperature of the plasma was way up and the number of fusion neutrons being produced was extraordinarily high. Instead of the usual 10 trillion or 100 trillion neutrons per second, the plasma was throwing off 1,000 trillion. Instead of the puny amount of stored energy typical of the "low mode," the plasma's energy was being contained much longer and more completely. The only factor of the fusion equation that was missing was high density of the plasma. By day's end, however, the physicists had no idea what had delivered these unanticipated results.

Rob Goldston, who had been running the "SNAP" computer analysis, lobbied for more time for Strachan and Mike Z., and the following day was granted.

Blind, empirical science took command. The researchers basically tried to do exactly what had been done the day before, uncertain about which of their steps was the key. By day's end, TFTR had produced another set of "supershots," and the following week Goldston led the team in trying to push the limits even further. Sensing real progress, Furth descended to the control room and tried his hand at running some experiments.

Within a few weeks the team had reached 5,000 trillion neutrons per second. Every day brought a new neutron record, but still the scientists had no firm idea of the reasons. As their apparent successes snowballed, the term "supershot" seemed ambiguous.

"'Super' changes from week to week," reported Kevin McGuire, a diagnostics specialist, at a TFTR meeting. "What was super two weeks ago is not super this week."

"It's like toothpaste," remarked Jim Sinnis. "Nothing is 'small' just 'large', 'extra large', 'family', 'massive.'"

At one meeting, "E-mode" for "enhanced confinement" was suggested. The "tiger mode," shouted someone else, recalling the Princeton University mascot. "Put a tiger in the tokamak!" But the "supershot" terminology could not be shaken, especially as the shots grew more and more super. By the end of July, the low density supershot plasmas were more than twice as hot as anything that had been seen before in a fusion experiment – 200 million degrees centigrade.

Harold Furth started preparing for a news conference. To be sure, TFTR had not achieved breakeven, but it was no longer a boring machine.

Looking awkward in suits and ties, the Princeton physicists congregated in the lab lobby outside the large auditorium where television news crews were setting up their equipment. Mike Z. was formal in a wool tweed jacket and tie, but he wore his sockless sandals as a kind of personal statement.

In the auditorium Harold Furth took his place at the podium facing several hundred people. The bearded professor recalled the long history of fusion research, how he himself had joined the effort in 1956 when the scientists needed temperatures 1,000 times better than what they were getting.

"Ten years later we hadn't progressed very much," he said, "and by a simple calculation I was able to see I would be 200 years old before we ever got in sight of the goal. But that is not how things developed. The new result on TFTR is that we have reached 200 million degrees and that is high enough so we are no longer just in the ballpark. We are at the temperature at which one can run the center of a reactor."

The key to the supershots seemed to be the cleaning of the tiled inner wall of the tokamak, which enabled the tiles to better absorb cool, unwanted particles from the plasma's edge. The balanced neutral beam injection seemed to help as well, Furth said.

The challenge now, he added, was to improve the confinement time and density of the plasma at these high temperatures and neutron levels. Furth described other, more subtle lessons from the recent experiments. There was reason to believe that a superhot plasma might not need to be primed with pulses of electricity to keep the reaction going, as the power industry had feared. In sum, said Furth, the scientific results should elevate the opinions of energy experts "about the real potential of fusion power."

Princeton's temperature record made front page news around the world. Even the China Daily carried an item. Telegrams and letters of congratulation rained on the Princeton lab, including a note from Roy Bickerton, the sardonic scientific manager at JET who wrote to Furth, "We are much encouraged by your results although you have given us a hard act to follow." Masaji Yoshikawa at Japan's JT-60 wrote his congratulations and added that "We at JT-60 will certainly try hard to match your efforts."

No sooner was the news out than Geoff Cordey, one of the experiment leaders on JET, jumped on the gossip line to Rob Goldston, asking for more details about the "recipe" for supershots. If TFTR could produce them, then its bigger European cousin should also have the capability. Goldston shared the recipe freely as Princeton had already reaped the public benefits of having been the first giant tokamak to break out of the low mode. A few days later, after a series of helium cleanup shots on JET, Cordey reported back that a brief test had already produced a 30 percent improvement in confinement time on JET and a twofold increase in neutron production.

"It just plain worked, which is quite delightful," said Goldston.

JET management, including the "mad" Rebut, later said it had high temperature, low density experiments on its agenda anyway, and that Princeton's temperature record, while interesting, was only a sideline on the trail to a reactor plasma that combined the requirements of temperature, density, and confinement time. A Plasma Olympics was looming in November in Kyoto, Japan. This would be another big showdown of the giant tokamaks, and JET apparently had some tricks up its sleeve. The race was still on.

Japan, already a world giant in technological applications and business, had never hosted the Plasma Olympics. So the selection of Kyoto as the

1986 site for the biennial international meetings was an acknowledgment of Japan's earned stature in the fusion community. The nation's JT-60 team was eager to produce grand results for the occasion. The meetings were customarily held in September, but the Japanese had lobbied for a postponement to November, offering the explanation that hotel space in Kyoto would be tight in September. The prevailing view among their international colleagues, however, was that the extension was meant to give the late-starting JT-60 extra experiment time to catch up with the other giant tokamaks.

The Japanese machine was housed in a plain, tan brick building set on a field of red-brown soil north of Tokyo, outside the small town of Naka. In 1985, JT-60 was the last entry in the giant tokamak competition, but its meticulous construction made up for any tardiness. The giant tokamak, similar in size to the Great Buddha of Nara and slightly larger than Princeton's TFTR, was painted in pristine white, its environment kept clean and free of debris. With amusement and reverence, the Japanese sometimes referred to JT-60 as "The Great Buddha of Naka." Employees were required to change from street shoes into dirt-free sneakers reserved for use in the control room or test cell. Visitors wore plastic bags around their own shoes.

Like a legion of postmen, the scientists and engineers in gray uniforms of short, zippered jackets and matching trousers populated the building's immaculate corridors. The workers wore color-coded hard hats signifying their employers – white for government workers, yellow for Hitachi, green for Toshiba. During the tokamak's initial phase, the companies responsible for building the machine kept large teams of workers on hand. Many employees of the research center lived in a bachelor dormitory on the center's grounds. Some married men stayed here during the week as well, reserving the weekend for their families. The fusion project was the center of their professional and social lives.

Although somewhat behind in international fusion competition, the Japanese were prepared to catch up. Lining one wall of the test cell was a complete set of the machine's neutral beam boxes, fully tested and guaranteed by the manufacturer, waiting for installation. JT-60's design also capitalized on trends that came too late for the TFTR, JET, and Soviet T-15 designs. The Japanese tokamak, for example, included a "divertor," a magnet configuration that could siphon off unwanted particles on the plasma's edge. However, for reasons of national politics, JT-60 was not equipped to use radioactive tritium. Without tritium it could never reach true breakeven, only a state of hydrogen fusion activity that could be pronounced "scientific breakeven." Had tritium been used, the Japanese would be able to say, the machine would have produced as much energy as it consumed.

The Japanese were also prepared to pursue fusion steadily and for the long haul. Their national energy plan included a commercial fusion reactor by year 2030, an acknowledgement that fusion would take generations to perfect. The formal dedication ceremony for JT-60 included two elementary school students from the Naka area chosen because they had been born seven years earlier, in the year of the machine's ground breaking.

Masaji Yoshikawa, considered the builder of JT-60, had the same proprietary feelings about the white giant as Don Grove at Princeton had about TFTR, and he believed his machine was a promising late entry in the race to breakeven. "We are behind JET and TFTR by two years," he acknowledged. "But we are in a good position for catching up. The way we build the machine is completely different. We make a contract and when we complete the contract the machine has been tested at full rating, so we can immediately begin operating at full rating if we want." Yoshikawa expected to crank up his machine and sprint into the high mode in time for the Plasma Olympics in Kyoto.

Like children bursting with a secret, the fusion scientists could not wait for the meetings in Japan to spill their charts and monographs. Two weeks before Kyoto, the American Physical Society's plasma physics division held its annual meeting in Baltimore. The Princeton physicists, a delegation from JET, and a surprise team that had worked on the Doublet machine operated by GA Technologies – formerly General Atomic – buried the low mode under a torrent of promising new results.

They told the gathering of several thousand colleagues that each of the machines, with its own distinctive style, had hit high temperatures and improved stored energy in their plasmas. Princeton had further "supershot" results. JET had a new alignment of magnetic fields that mimicked a divertor. Doublet, which had just been enlarged to a chamber bigger than TFTR's, had the advantage of both a D-shaped plasma and a genuine divertor.

Mike Zarnstorff delivered a self-effacing report for Princeton. The supershots were discovered "entirely by accident," he said.

Roy Bickerton of JET followed. "It's nice to hear that accidents can have a positive effect," he deadpanned, to gales of laughter. A recent water line break at JET had temporarily turned that vacuum vessel into a goldfish bowl. "On JET we've had 500 liters of water in the torus and it didn't produce anything positive," Bickerton added. Nevertheless, JET researchers reported that with the new magnetic configuration JET had reached temperatures of up to 125 million degrees centigrade with low-density plasmas. Confinement time and stored energy had also improved. Though problems remained, JET had definitely left the low mode behind.

The exploration of the new mode had just begun, and the physicists were euphoric. In the past few years, meetings of the American Physical Society had been depressing affairs, a mere marking of time. Now the big machines were starting to hit pay dirt and it electrified scientists at the leading labs and universities. An atmosphere of great hope suffused the meeting. Even more encouraging was news delivered by John Clarke, head of the Department of Energy's Office of Fusion Energy. In a follow up to the fusion rumblings at Geneva, President Reagan had just issued a "Presidential Initiative," a formal order, directing the agency to study an international fusion project that would include the Soviet Union and perhaps Europe and Japan. "We've never had an endorsement on such a high level," said Clarke.

Meanwhile, Clarke added, the United States was about to sign a new bilateral agreement with the European Economic Community to cooperate in fusion research. It was a step closer than exchanges with individual countries. And it made the idea of an internationally financed and operated engineering test reactor more realistic.

"We need to get Europe and Japan to look beyond themselves," said Clarke. "Everyone is saying 'I want to do it myself, first.'"

Meanwhile, the Japanese had kept silent. The Great Buddha of Naka was stuck in the low mode.

JT-60's power systems and computers had worked beautifully, but it could not break into the high mode, and no one knew why. Even more galling to the Japanese managers was the fact that one of their small research tokamaks, housed in another lab in Naka, had suddenly stumbled into the high mode without the benefit of divertors, D-shaped plasmas, or any of the hocus-pocus the giant tokamaks were using. Rob Goldston and Mike Zarnstorff checked out the little tokamak's data and agreed that the machine produced good results no matter how it was run – high density, low density, you name it. The challenge was to figure out why.

Despite this challenge, or perhaps because of it, Kyoto proved to be "basically a love fest for the tokamak," Goldston concluded.

On the political front, there were other hopeful signs for the fusion community that emerged from the meetings in Japan. Bruno Coppi of MIT had noted with surprise that during a set of speeches marking the twenty-fifth anniversary of Nagoya University's Institute of Plasma Physics, a Soviet speaker had mentioned Andrei Sakharov as one of the founders of the Soviet Union's Lebedev Physical Institute, also twenty-five years old. The written material distributed with the speech also listed Sakharov. Coppi had not heard Sakharov's name uttered by a Soviet scientist in a public forum since the banishment to Gorky. Something was up, and when he returned to Boston he informed Sakharov's relatives there.

Within a month, Coppi saw that he had read the tea leaves correctly. The wishes of Gorbachev for decreased military spending and better relations with the West had turned fortune in Sakharov's favor. On December 23, 1986, after nearly seven years of banishment, Sakharov returned to Moscow with his wife, Yelena Bonner. That same day, trailed by an entourage of jostling camera crews, he went straight to the Lebedev Physical Institute, his old lab, to attend a scientific seminar. He was greeted there with applause. When asked by Western reporters about his plans, he said he hoped to return to work in fusion. He told *Newsweek:*

My main job – what will take up a lot of time – will be my scientific work in the area of physics of elementary particles, in the areas of cosmology – that's what I am interested in most of all – and also I want to participate in the discussion of thermonuclear problems and the peaceful use of thermonuclear energy, which I began to deal with in 1950. I have not worked on it for a long time and now I want to work on it again.[1]

At Princeton, Harold Furth immediately sent a warm message to Sakharov. "We were delighted to learn of plans for your return to Moscow," Furth said in a telegram. "The first thirty-five years of tokamak research have been a great success. We hope to have the opportunity to tell you about our recent work." Furth also shot off a telegram to his friend Velikhov. "We were all happy to hear the news about Andrei Sakharov. This is a most constructive development for international science, and particularly for collaboration in fusion research."

In Boston, Coppi was thrilled by the news – and by the measure of relief it would bring to tensions within the international fusion community. "This was a thorn in our flesh," he said.

Over the next few months, Sakharov had encouraging words for the efforts of Mikhail Gorbachev to loosen the Communist Party's tight controls over Soviet society. And quickly the scientist returned to the political arena, eventually winning election to the new Soviet legislature, the Congress of People's Deputies. If fusion had been on Sakharov's mind, it soon took a back seat to his participation in the historic political changes being wrought under Gorbachev.

Shortly after Sakharov's release, scientific representatives from the governments of the United States, the Soviet Union, the European Economic Community, and Japan sat down together in Vienna. It was agreed that the four would together design a fusion test machine called the International Thermonuclear Experimental Reactor (ITER). Its purpose would be to prove the engineering feasibility of fusion as an electric power source. Such a machine was already part of long-range fusion plans for all four governmental groups and could cost up to $4 billion. A decision to actually build the machine jointly would wait for completion of the conceptual design.

Government leaders in other times had found political utility in the idea of fusion – recall that the science had been used by the United States to steal the show from the Soviets at the 1958 Atoms for Peace conference. European leaders had also used fusion to demonstrate the viability of a European Economic Community project. Now the superpowers had found common ground in fusion as a vehicle for expressing good will and cooperation at a time when the Soviet Union in particular was pushing for arms control agreements.

The idea of a truly international reactor had been limping along since 1979 under the auspices of the International Atomic Energy Agency. During that decade, planning studies had been turned out by a skeleton crew of forty scientists from the four big fusion powers during what were termed "workshops" in temporary offices in Vienna. The delegates had dutifully served time in Vienna, always wondering if their work would ever come to anything or was simply an exercise in good will. With the agreement to design ITER, the scientists had reason to hope.

The ITER design team was given a permanent base at the Max Planck Institute of Physics at Garching near Munich, West Germany. It was alloted three years for the assignment and a budget of $170 million, contributed equally by the participants.

"There's a gleam in everyone's eye and a hope that after three years the political climate will be warm enough to build it," said Manfred Leiser, head of the physics unit of the International Atomic Energy Agency, which orchestrated the four-way agreement.

Bit by bit, the scientific difficulties and the expense of fusion research were accomplishing a small political miracle.

The scientific and political advances were some of the most satisfying in fusion's grinding trek toward commercial viability. Even so, in the late 1980s, the international fusion community arrived at a crucial and troublesome turning point. Its trained corps of plasma physicists and nuclear engineers would be without a new generation of test reactor if construction did not start soon on a pioneering experimental machine that could be up and running in the early 1990s.

"The European program is approaching a precipice," explained Roy

Bickerton, as he and a forty-year-old colleague, Enzo Lazzaro, sat discussing the dilemma facing the younger generation of plasma physicists. Bickerton himself was close to retirement. The European fusion plan projected work on JET to the year 1992. After that the schedule was blank. The next machine, Bickerton pointed out, would take seven years or so to build. Unless construction started immediately his younger colleagues would have no machine to work with for a few years. His fear was that precious knowledge would be lost, specifically the transmission of observations and techniques from one group to another that takes place through personal interaction, not by reading scientific papers. There would be a terrible gap.

"I think it's awful – this very careful approach," Lazzaro said to him. "What would we lose if we started now a more daring approach and shorten this time? Who is going to devote his entire life to a cause that doesn't even promise to converge in his lifetime?" The next machine, Lazzaro said firmly, "must be decided soon."

Researchers at Princeton were feeling the same tug of time. For the health of the Princeton lab, the scientists felt a new machine was imperative. The U.S. Congress had already turned down their bid for a large machine to take the fusion experiments to ignition, but they persisted. In the budget-conscious atmosphere of 1987, a national team of American designers devised a cheap, compromise machine whose scientific value could be defended. It would be a $357 million ignition tokamak, much smaller than Princeton's TFTR, that for economy's sake could be linked up to the giant tokamak's already existing power systems and computers as soon as the big machine was shut down. With intense, compact magnets, of the sort Bruno Coppi and MIT had been advocating for a decade, it could bring forth a self-sustaining though short-lived plasma, useful for examining ignition physics but not of the greatest use for designing full-scale reactors.

In Europe, Paul-Henri Rebut was pushing for the community to think big. He wanted a mammoth ignition device, "a real furnace," that could burn a self-sustaining plasma producing 1,000 megawatts of power over several hours.

In 1987, during the hopeful atmosphere following the supershot successes, President Reagan's Department of Energy endorsed the Compact Ignition Tokamak to be built at Princeton. But it was a mixed message. The entire fusion program suffered further budget cuts over the next two years, forcing researcher layoffs at Princeton and the conclusion of experiments at smaller labs. Construction of the compact tokamak was put off until the late 1990s.

With its decision to push for the economical ignition machine, the politicians were presuming an imminent breakeven success on Princeton's TFTR or at one of the world's other giant tokamaks. But breakeven was not to come.

Although the Princeton plasmas seared up to 300 million degrees centigrade by 1988, their energy output was meager by reactor standards. Such hot plasmas were not very dense and did not last more than five-tenths of a second before escaping the magnetic bottle. Regarding breakeven – where energy output equals energy input – the best Furth could say was that the plasma was making about a third as much energy as it was consuming. In the most optimistic terms then, had the machine been fed radioactive tritium fuel, output would have reached only about half of input.

The JET directors said they were getting a similar energy output from their best plasmas. JET's plasmas were not usually as hot as Princeton's, but they had a very long energy confinement time, up to 1.4 seconds. JET's directors had applied to the European Community for a four-year extension on their experiment to 1996. Without a better performance it would be nearly impossible to get the European Community to commit itself to the next machine, leaving hundreds of plasma physicists and engineers stranded between tokamaks.

Success was relative. "The plasmas we make nowadays were only dreams of the people in the mid-1970s," said Princeton's Dale Meade. Yet as far as the researchers had come since the 1950s in producing plasma temperatures, densities, and confinement times, they would have to go many times further still to get to breakeven. The years were slipping by and government interest was waning in the face of other, more pressing priorities. Yes, breakeven was in sight, but to get there was still problematic.

14
Fusion's past and future

From the ancient Greek legend of Icarus, we know that for millenia man yearned to fly before actually discovering the means to do so. Nature had provided the tantalizing example of the bird and left it to human ambition and intelligence to eventually find the solution. That mankind would some day fly was certain.

During human history, he has also admired and depended upon another of nature's marvels, the sun. It provides him with constant, reliable, limitless energy in forms that include life-sustaining light and heat. Only in this century, however, has man deciphered how the sun works. With that understanding was born the ambition to *make* a sun. The tantalizing example of endless energy greets him each morning. That man will some day create his own sun on earth seems certain.

The force of this vision, like the yearning to fly, is almost primal, driving individuals and governments to paroxysms of excitement any time the goal has seemed within reach. Even now, after decades of disappointments, the world body politic still resonates should someone, yet again, cry "Eureka, fusion is at hand."

In the spring of 1989, that was exactly what happened when a bizarre fusion experiment at the University of Utah convulsed the world's physics establishment in a manner not seen since Juan Peron's equally strange announcement in 1951 that Argentina had built a working thermonuclear reactor.

Two little known chemists called a news conference at the university and shocked fellow scientists and a crowd of reporters by announcing that they had created fusion at room temperature in a glass jar filled with water. They called the method "cold fusion" and said they had done it with equipment available in any college laboratory.

Thus did B. Stanley Pons, chairman of Utah's chemistry department, and Martin Fleischmann, his former professor at the University of Southampton in England, set in motion a two-month frenzy of experimentation at laboratories around the globe as other researchers sought to validate or disprove their incredible claim. This was hampered by the fact that the Utah team, apparently worried about patent rights, had revealed only the barest details of its experiment. Violating accepted scientific protocol, they had announced their results in front of television cameras instead of through a paper in a scientific journal.

Although scientists at major laboratories reacted with much skepticism, there was still a measure of suppressed excitement – the taming of fusion would be an historic event. Meanwhile, newspapers, magazines, and television exploded with the news that the solution to the energy crisis had been found seemingly in "fusion in a jar." Cold fusion was featured on the covers of the two largest weekly news magazines in the United States and was the top item on network television news shows. Fusion had never received such prominence.

Fleischmann and Pons described how they had produced energy by passing an electric current through two electrodes immersed in a bath of "heavy" water formed with deuterium, the hydrogen isotope. They said the key was that one electrode was made of palladium, a metal that has the unusual property of absorbing great quantities of hydrogen gas.

Sending an electric current through the device, they eventually observed an emission of energy in the form of heat. They said the experiment generated four watts of energy for every watt of electricity used. Finding no chemical explanation for the result, they deduced that nuclear fusion was responsible. They postulated that the hydrogen nuclei, driven together in the latticelike structure of the palladium, were able to come close enough to fuse.

With this scant description to go on, physicists by the thousands laid aside their current projects and slapped together their own cold fusion experiments. They based their equipment on details gleaned from purloined copies of an unpublished Fleischmann and Pons paper and by enlarging photographs of the apparatus taken by news reporters.

B. Stanley Pons, University of Utah chemistry professor who, with Martin Fleischmann, in 1989 reported creating "cold fusion" in a bottle at room temperature. Courtesy University of Utah.

"If this thing is what they think it is, it's better than the gold rush."—UTAH GOVERNOR NORMAN H. BANGERTER.

Teams of physicists were set to work around the clock. Palladium prices jumped from $145 to $185 per troy ounce just after the Utah announcement.[1]

The table top fusion scheme – financed by $100,000 from Fleischmann and Pons' own pockets – was an open challenge to the way fusion had been researched for more than thirty years. If it proved to be true, it would make a mockery of the billion-dollar tokamak experiments and a tragedy of the lifetimes spent on "hot" fusion. A Nobel Prize for Fleischmann and Pons would surely follow as well as riches from patent claims.

The idea for their simple experiment had come to them on a hike up a mountain – apparently a fruitful environment for fusion thoughts. Lyman Spitzer had conjured up the Stellarator while traversing a Colorado mountain some four decades earlier.

Even as they tried to duplicate the Utah results, conventional fusion researchers scoffed at the experiment. If fusion were responsible for the excess energy, they said, Fleischmann and Pons should have been dead – killed by a stream of trillions of neutrons per second. In fact, the two researchers had failed to measure neutron emission, a characteristic sign of fusion as it was understood by plasma physicists. The Utah team speculated that a previously unknown form of fusion bonding was taking place, one that did not emit so many neutrons.

Still, reports began to come in daily from labs that had partially reproduced the chemists' results. The debate over cold fusion became a spectacle of "science by press conference" played out in a rush for a public audience instead of in the usual measured pace of scientific exchange and peer review.

Texas A & M University gave the Utah team a big boost when it announced that its version of the experiment had produced up to 80 percent more energy that it consumed. At the Georgia Institute of Technology, scientists announced they had detected neutrons and tritium, a radioactive form of hydrogen, in a similar experiment.

Also laying claim to partial cold fusion success were labs from Italy to Czechoslovakia to India and Brazil. Peter Hagelstein, a famous MIT theoretical physicist, said cosmic rays were responsible for touching off the reaction and filed his own patents. It seemed no one wanted to be left out of the action.

The state government of Utah, which could taste wealth and glory in the air, was particularly excited. "If this thing is what they think it is," said the state's governor, Norman H. Bangerter, "it's better than the gold rush." The state legislature went into emergency session to vote $5 million in seed money for commercial applications of fusion technology. Industry representatives descended on the university, lining up for development rights. The state attorney general's office filed five patent applications on behalf of the University of Utah to protect any commercial use of the technology. At a hastily called congressional hearing on cold fusion, Chase N. Peterson, president of the University of Utah, asked for $25 million in federal funds to create a cold fusion research center.

It had been University of Utah lawyers in fact who pressed Fleischmann and Pons to announce their discovery ahead of publication as a way of nailing down their historic and valuable claim. The primary factor in the rush was that, by strange coincidence, Steven Jones, a researcher at state rival Brigham Young University, also had been working on cold fusion schemes for several years, although his experiments had produced no excess energy. He had found out about the

University of Utah work only recently and had suggested a collaboration. He intended to publish his results within weeks. At a meeting of officials from both schools, the two teams could only agree to simultaneously submit papers to the prestigious British journal *Nature,* setting March 24 as the date and to avoid speaking publicly about their work until that time.[2]

On March 23, however, the University of Utah called for a Fleischmann-Pons press conference without notice to Jones. The Brigham Young researcher felt he had been hoodwinked. The University of Utah defended its decision to go public, saying news of the results was already spreading.[3]

Besides the rivalry between the University of Utah and Brigham Young, the debate over the validity of cold fusion revealed other antagonisms between chemists and physicists, between Easterners and Westerners. At a meeting of the American Chemical Society shortly after the cold fusion announcement, 7,000 fellow chemists gave the Utah team sustained applause and suggested that the physicists were just angry that chemists had invaded their territory – the heart of the atom. Utah politicians suggested that an East Coast fusion establishment had a vested interest in disparaging the cold fusion discovery.

But the criticisms continued. Cold fusion suffered a blow when the two American labs that initially ratified Fleischmann and Pons' results had to issue retractions. Texas A & M University, which had detected excess heat, eventually found the measurement was false, due to an improperly grounded wire. Even worse, a palladium sample was found to be contaminated by tritium, falsifying results. Georgia Tech, which had detected excess neutrons, revealed that its neutron counter, affected by the warmth of the water in the fusion jar, was giving inflated readings.

None of the large laboratories best equipped to verify the experiment could produce results such as the Utah team described in a paper eventually published in the Swiss periodical *Journal of Electroanalytical Chemistry* and *Interfacial Electrochemistry*. Exhaustive collaborative experiments by Brookhaven National Laboratory and Yale University came up empty. Livermore national laboratory and Great Britain's Harwell Atomic Energy Research Establishment failed to generate cold fusion according to their meticulous measurements of heat, neutrons, and other radiations.

At an American Physical Society meeting in Baltimore, physicists presented forty papers critical of the cold fusion claim and assembled a panel of experts that lambasted the Utah team. The crowd applauded when Dr. Steven E. Koonin of the California Institute of Technology in Pasadena said the phenomenon was a result of "the incompetence and perhaps delusion of Pons and Fleischmann."[4] MIT researchers suggested how the Utah team had misinterpreted the data from which they inferred the rate of neutron production in their experiment. If neutrons were present at all, said MIT, it was at a level far below what was reported.

"Unfortunately, a lot of time and effort has been wasted due to this blunder," said Ronald R. Parker, director of the MIT Plasma Fusion Center.[5] The standing of Fleischmann and Pons was further eroded when they withdrew a paper they had finally submitted to the journal *Nature*. The magazine had asked for an amplification of the experiment before it could publish the paper. The two scientists said they were too busy to make the requested additions, according to the journal.[6] In an editorial a few weeks later, *Nature* editor John Maddox wrote that the Utah team's claim, "is literally unsupported by the evidence, could be an artifact [due to something completely unrelated to the experiment] and, given its improbability, is most likely to be one." He said the team had announced its findings before doing any control experiments to verify results. That was an "astonishing oversight" and a "glaring lapse from accepted practice."[7]

All along, Fleischmann and Pons contended that other teams were not performing the experiments correctly. They said, in fact, that their latest experiments showed energy production 100 times greater than the electrical input. To help put an end to the imbroglio, they said they would work directly with Los Alamos national laboratory to set up a duplicate experiment. But eventually the University of Utah put off that agreement for the sake of protecting patent claims.

The cold fusion episode ended with the hot fusion establishment convinced that the Fleischmann-Pons claim of net power-producing fusion was spurious. The fraternity withheld judgment on Jones' more modest claims of occasional fusion reactions in certain metals. Jones himself told a Los Alamos-sponsored meeting on the subject that, because the neutron rates he was detecting were so low, "From the work I've done, I would say cold fusion is worthless as a source of energy."[8]

No new rival was poised to destroy the fusion world as the plasma physicists knew it. The burden of proof was left with the Utah scientists, and the fusion fraternity went back to its tokamaks. Once again in fusion's odd history, respected scientists had been drawn into premature declarations of success by fusion's utopian allure and all that it would mean to mankind.

It would have been grand if, in the year 1989, fusion had been solved in a glass bottle. But fusion research, it seemed, was destined to move at a snail's pace into a distant future, the plasma captured in a magnetic bottle of dubious quality.

In the end, the Utah claim had done the field some good, for now millions of people who had never heard of fusion had been exposed to the dream. Fusion had been at the center of world attention for two months of excitement. A 200 million degree plasma at Princeton had generated only one day's attention. That was worth something.

Fleischmann and Pons had shown one thing. The shimmering dream of fusion energy could still electrify the imagination. At a moment's notice, the world would be ready to embrace a fusion future.

Nearly forty years after Lyman Spitzer took the ski lift up the mountain in Aspen and invented the Stellarator, the fact remains that the world does not have a power-producing fusion reactor. What's more, although the research has come a long way, from plasmas of several hundred degrees to 300 million degrees, this grand project will still require another generation or more to complete.

Why has fusion taken so long? Would more money and support have made a difference? Where did the scientists go wrong? What lies ahead?

As four decades of research and machine building have made perfectly clear, controlling fusion is an extraordinarily difficult scientific problem, more difficult than anyone had imagined in the beginning. In addition, it is an overwhelming technical and practical problem for which theory has offered little help.

In the ideal world of Spitzer's first imaginings, the plasma was confined in an infinitely long and constant magnetic field. Closing that field into a real-world shape transformed the problem into a wrestling match with machinery and currents of great complexity. What in

Spitzer's theories could account for the cracking and peeling of tiles along the inside of the machine, sending "boulders" of carbon cascading into the pure plasma? What was the real configuration of the magnetic fields as the current fluctuated? Did the plasma move in a predictable way or was it moving chaotically as one theory proposed? The only way for the physicists to learn how to master the plasma was by trial and error on ever bigger machines. The only concrete rule was that bigger so far had always proved to be better.

Looking back, it is unfortunate that years were frittered away in efforts to define plasma behavior *before* building new machines. Such behavior could have been discovered raw in speculative devices. An almost unscientific kind of political courage was lacking. That problem remains today as governments pause on the edge of commitment to the next $4 billion device. Only by building on such a grand scale will the world ever determine whether a magnetic bottle is the way to hold a fusion furnace.

Sebastian Pease, who participated in virtually all of Britain's major fusion work, has come around to the view that the scientists should have built bigger sooner despite the unsure grasp they had on plasma behavior. In 1959, Pease had won fame on the infamous Zeta machine. He then built a statesman's reputation by orchestrating the verification of Artsimovich's tokamak results and helping to bring about the European JET project. In 1986, as he prepared to retire from the field, he had this to say about what might have been done differently:

"If you ask me that question, I will tell you that we didn't have the courage to persist with the large apparatus," he said. The British scientists of Zeta's time "came under really quite heavy fire" for spending such large sums of money when the scientific data was so difficult to gather and interpret. Instead, the scientists were encouraged to understand plasma physics first, before building any more large machines.

"We in England spent a great deal of time agonizing over what the successor experiment to Zeta should be, and in the end nothing was built. But we should simply have built the simplest thing on a fairly large scale, and if we had done so, it would have been very close to JET."

More vigorous financial and social support would have helped fusion, but really, an evolution of how scientists think about fusion is what is really needed. There is a "human time scale," said Pease, set to the rhythm of how ideas evolve and how long it takes people "to actually get round to what is quite obvious in the end. Maybe you have to wait for the right genius to turn up in order to clarify people's thinking. Why didn't people know about the law of gravity before Galileo? Well, I think it was not a matter of not having enough money to support them. It was a matter of evolution of human thinking."

Derek Robinson, the Englishman who spent the first year of his marriage in Moscow getting an ulcer and working on Artsimovich's tokamak, is still an optimist about the tokamak. His latest project is a small and versatile research machine that can be converted from tokamak to Stellarator at will. He named it "Compass," he said, because he hoped it would point the way to the fusion reactor of the future.

Like Pease, Robinson also wishes bigger machines had been attempted sooner:

If you look back, we had operating in this country in 1957 the very large Zeta device, a device which is considerable in scale and can carry currents certainly comparable to those on TFTR and JET. In other words, we knew engineeringwise, technologywise and in a very rough sense physicswise what to do. We built such a device. But when we built it we didn't really know what to make of it when we actually produced our plasmas in there. We were puzzled by the whole range of physical phenomenon which it showed.

It has taken us perhaps twenty or more years to realize actually what those people were looking at on that device. When we made those measurements, the way we looked at things, the way we analyzed them, were very crude and very simple. It was some time later, more than ten years, before we really learned how to measure what was in the middle of that hot plasma. What have we done? Was it really hot?

But in principle, the scientists knew how to make a fusion machine. Robinson finds remarkable the patent filed in 1946 by Sir George Paget Thomson and Moses Blackman for a ring-shaped thermonuclear device, stating:

People then knew exactly what to do if you wanted to translate fusion to a power-producing system. Nobody built anything or even really thought much about it but there, sitting there, is exactly the embodiment of what we've got now. You might say, why is it thirty, forty years since Thomson put that in? Why have we now, only now, got such a system? It's really because the global understanding was there, but the detailed one was not.

A detailed understanding of the fusion process in a machine would have to wait for new technologies, said Robinson, and with them, new tools to chart and verify the plasma's behavior.

In 1985, after the sudden death from a heart attack of Hans-Otto Wuster, the Frenchman Rebut became director of his beloved JET. Looking back and looking forward, Rebut berated his fellow plasma physicists for hesitating to build big and bold.

"At each step the knowledge was not sufficient to have the courage to go ahead," he conceded. "But there is a general tendency not to be harsh enough in this field and to go too slowly, not to make the necessary step large enough." Thinking too small, said Rebut, invites its own risks because the safety margin to reach a certain goal may be too short. Rebut continued:

This is somehow a lack of courage from the fusion community, too afraid about the size, too afraid about the capital money without realizing that by doing so the whole cost is at risk. It's all around the world; I'm not only speaking of the States. It's Europe too. Somehow it's a lack of confidence of some people with maybe a too small view of the future.

I believe there is little doubt that we – I mean the fusion community in a very large sense – will be able to build a reactor from a physics, technological point of view, and this confidence was not here ten years ago.

Rebut, of course, has always been an optimist about fusion. "I must say, at a certain time I was among the few who were believing that fusion could be a reality."

Before his death on December 14, 1989, Andrei Sakharov came to believe that fusion's time as an historic power source had passed. In October 1988, he told an interviewer from New Jersey Network television

I think that controlled fusion reactors will decide problems of energy production which already have other solutions. I do not think fusion will change the world. It's purpose does not have that revolutionary character that the other great discoveries of the twentieth century have had, when something was being solved for the first time.

All the same, I think that practically it will be very important for humanity. I think that large-scale atomic energy production will receive a great deal from the first stage of controlled fusion. I think that this will take the form of the breeder reactor – controlled fusion reactions to obtain uranium for use in atomic energy production.

Edward Teller had always said that breeders would be the first and logical use of fusion reactors. But, in 1987, at the age of seventy-nine he still saw that time as distant. "I hope very much that the process of controlled fusion will become practical at some point in time, but . . . I do not expect that that time will come during my lifetime."[9]

Lyman Spitzer entered fusion as a believer and ended up as an agnostic. Interviewed in his office at Princeton University at age seventy-two, he was still churning out ideas about the physical world – and still climbing mountains and skiing on his vacations. What could have been done differently? he was asked.

He replied:

When I see what's happening now at the [Princeton] plasma physics laboratory, I sometimes ask myself what would have happened if we had built some enormous machine of the sort they are now building. I think we would have achieved much earlier the results that they are now achieving. But in point of fact, it would have been difficult to raise the money for that sort of thing or to have had the confidence we needed to take such a giant step.

To get to fusion's future, he continued:

First of all they have to demonstrate technical feasibility which they really haven't yet done. They're getting close, but they're not really there. And who knows, when you're climbing a mountain, until you're out on the summit, you're not really certain you're going to make it. It may look fairly straightforward, but it seems to me it'll be a few more years before they've demonstrated that the laws of physics pose no barrier to the construction of an effective, economic fusion reactor.

Spitzer was cold and precise in his assessment of fusion's potential. He said he had "never been positive" that fusion's problems could be solved and "I'm not positive now. I've always felt there was an appreciable probability that an economic reactor – a reactor that would generate more power than it consumed – could be built. Only a probability, not a certainty. One's assessment of that probability now would be higher." Then he added, in his characteristically understated way, "A fifty percent probability of getting a power source that would last a billion years is worth a great deal of enthusiasm."

The men in search of endless energy are still searching after all these years. As Spitzer says, their search "is worth a great deal of enthusiasm." The world's plasma physicists have not delivered a working fusion reactor as promised, but their success seems inevitable if they are given the machinery to explore the plasma. If they build a behemoth that is at first of no commercial use, certainly the refinement and the scaling down can come later, after the plasma has been brought to its knees. However, for an individual nation to build such a machine would be financially impossible. Here, the imperative is more political than scientific: international collaboration. An agreement to design such a machine – the ITER – has already been signed by the European Community, United States, Soviet Union, and Japan. That design will only be taken seriously if governments also agree now to build a machine together when the design work is finished. At present, the partners are free to go their own way when the blueprints are done.

The cooperative atmosphere of Gorbachev's glasnost and the consolidation of the European Community into one market may help advance the idea of a collaborative project. The growing popular power of the environmental movement – as individuals wake up to the peril of global warming and the depletion of the earth's resources – may also encourage the construction of a fusion test reactor.

An act of political courage and foresight – a concrete commitment to build ITER – is needed now, not when the paper study of ITER is completed and gathering dust while the world chokes on its own oil-burning exhausts.

In the meantime, the fusion fraternity has grown old, its pioneers stepping one by one into retirement, its middle aged enthusiasts moving into the shrinking ranks of management while the crop of young recruits dwindles due to fusion's uncertain future.

In five years of observing the fraternity, I have witnessed the ebb and flow of its hope – the high moments of astronomical temperature records, the low moments when governments betrayed fusion's future with the budget scythe. Despite the fluctuations in government support for the research, I have come to feel that the fellowship has largely itself to criticize for the unmet promise of fusion. The scientists, as Bob Hirsch has suggested, have gotten lost in their daily work, in the interminable analysis of data trends, in the romantic dance with the elusive plasma, fascinating, tantalizing, ready each day with a new set of movements to learn. The fraternity must look up from its data charts to the future, be willing to set aside esoteric work that may be useful in refining a reactor someday, and move on, despite a poor understanding of plasma, with unscientific boldness to machines that look like reactors.

Moreover, the fraternity must speak with one voice, from self-depreciating MIT to brassy Princeton, to the dapper JET and the ancient Kurchatov laboratory, across to the polished Japanese atomic energy institute. That voice must say that controlled fusion energy is imperative and that the way to have a reactor is to build one, now. The Wright brothers started their research with an airplane that they intended to fly, not one that would theoretically fly were wings and fuel added. Without a bold attitude fusion will remain a mirage, always out of reach, forever only twenty years away.

Notes

PROLOGUE

1 R.F. Post, "Prometheus Updated: The Fusion Quest," 30 April 1981 UCRL–85876. Courtesy Lawrence Livermore Laboratory.
2 R.F. Post "Controlled Fusion Research—An Application of the Physics of High Temperature Plasmas," *Reviews of Modern Physics,* 28: July 1956, 338–362.
3 Christopher Flavin, "World Oil: Coping With the Dangers of Success," *Worldwatch Institute,* July 1985, p. 25. See also John S. Herrington, U.S. Secretary of Energy, speech to the Commonwealth Club, San Francisco, California, September 13, 1985.
4 Ruth Leger Sivard, "World Energy Survey," 1981.
5 Claire E. Max, "Uranium Resources and the Development of Fission and Fusion Breeder Reactors," *Energy and Technology Review,* October 1984.
6 "Beyond Oil: The Threat to Food and Fuel in the Coming Decades," Complex Systems Research Center, University of New Hampshire, 1985.
7 Joseph D. Parent, "A Survey of United States and Total World Production, Proved Reserves, and Remaining Recoverable Resources of Fossil Fuels and Uranium as of December 31, 1981," Institute of Gas Technology, 1983.

CHAPTER 1

1 *New York Times,* 25 March 1951, p. 8.
2 Lyman Spitzer, Jr., Project Matterhorn papers, PPL, boxes 326, 327, Princeton University Archives, Princeton, NJ. Following account from Spitzer papers, author's interview with Spitzer, and other members of the Matterhorn team. See also from Earl C. Tanner, "Project Matterhorn, 1951–1961," Princeton University Plasma Physics Laboratory.
3 Harold M. Mott-Smith, Memoir May 1967, 11–12, General Electric Research Laboratory Archives. (Courtesy of George Wise.)
4 Edward Teller, "Peaceful Uses of Fusion," speech at the Second United Nations Conference on the Peaceful Uses of Atomic Energy, September 1958.

5 The Hiroshima atomic bomb was a 20-kiloton device that, in order to reach critical mass, was comprised of 45 kilograms of uranium. Only a small portion of the fuel actually fissioned, an estimated one gram. The exact amount is still classified secret by the U.S. government.

6 Edward Teller, Energy From Heaven and Earth, W.H. Freeman, San Francisco, 1979, p. 202.

7 The U.S. launched into space just such a device, dubbed Hubble, in 1990.

CHAPTER 2

1 I.V. Kurchatov, "On the Possibilities of Producing Thermonuclear Reactions in a Gas Discharge" (Lecture at Harwell, 25 April 1956. See Culham Library reference copy).

2 A.D. Sakharov, "Theory of Magnetic Thermonuclear Reactors," Part II, AN SSSR, *Fizika Plazmy i Problema Upravliaemykh Termoiadernikh Reaktsii.* See also I.E. Tamm, "Theory of Magnetic Thermonuclear Reactors," Part I. 1950 (58).

3 Details on fusion decision drawn from I.N. Golovin, "I.V. Kurchatov, A Socialist-Realist Biography of the Soviet Nuclear Scientist" trans. William H. Dougherty, 1968. Also author's personal correspondence from I.N. Golovin, 31 October 1986.

4 A.S. Eddington, "The internal constitution of the stars" (report of the British Association for the Advancement of Science, 1920).

5 R. Atkinson and F.G. Houtermans, "Zur Frage der Aufbaumoglichkeit der Elemente in Sternen," *Zeitschrift fur Physik,* 54 (1929).

6 This section draws in part from a detailed discussion of early British fusion science by John Hendry, "The Scientific Origins of Controlled Fusion Technology," *Annals of Science,* 44 (1987):143–168.

7 G.P. Thomson archive, Trinity College Library, Cambridge, folder E88(2). See also R. Carruthers, "The Beginning of Fusion at Harwell," *Plasma Physics and Controlled Fusion,* 30, No. 14, December 1988. See also Paper TC27, Jan. 23, 1947, UKAEA file 330/10/16/3pt1, courtesy UK Public Record Office AB16/727.

8 Details on Thonemann and Tuck taken in part from Hendry, Joan Bromberg, "Fusion: Science, Politics and the Invention of a New Energy Source," 1982 and Carruthers. Also personal correspondence to the author from Thonemann, 19 May 1989.

9 Thonemann, personal correspondence with author 11 July 1989.

10 Thonemann, 11 July 1989.

11 Golovin, p. 78.

12 *New York Times,* 8 October 1957.

13 Lewis L. Strauss, Chairman, U.S. Atomic Energy Commission. Remarks prepared for delivery to Atomic Industrial Forum—American Nuclear Society, Waldorf-Astoria Hotel, New York, 29 October 1957. From U.S. DOE Archives, 326 U.S. Atomic Energy Commission Collection Speeches Box 1 Folder Lewis Strauss.

14 Proposed new Sherwood Classification Guide from AEC, July 5, 1957. Folder: Geneva Conference 1958. Spitzer papers, Princeton University.

15 Memo from A.E. Ruark, Chief, Controlled Thermonuclear Branch, Division of Research USAEC to Tuck LASL, Spitzer, York UCRL re Sherwood Publication in

the period from early February to Geneva, 26 December 1957. Spitzer papers, Princeton University.

16 *New York Times,* 25 January 1958.

CHAPTER 3

1 Lewis Strauss, *Men and Decisions,* p. 367.

2 L.A. Artsimovich, "Research on Controlled Thermonuclear Reactions in the U.S.S.R." United Nations Second Conference on the Peaceful Uses of Atomic Energy. Session 4 P/2298, September 1958.

3 *New York Times,* 3 September 1958, p. 9.

4 Edward Teller, "Peaceful Uses of Fusion," United Nations Second Conference on the Peaceful Uses of Atomic Energy, Session 4 P/2410, September 1958.

CHAPTER 4

1 Harold P. Furth and Richard F. Post, "Advanced Research in Controlled Fusion," 10 December 1964, UCRL–12234, p. 4. Courtesy Lawrence Livermore Laboratory.

CHAPTER 5

1 L.A. Artsimovich, "Experiments in Tokamak Devices," August 1968. International Atomic Energy Agency meeting Novosibirsk.

CHAPTER 6

1 For a detailed discussion of the relative safety of fusion vs. fission reactors see R. Hancox and W. Redpath, "Fusion Reactors—Safety and Environmental Impact," 1985 (CLM-P750). Culham Laboratory, Abingdon, Oxfordshire, England.

2 *New York Times,* 20 July 1971.

3 *New York Times,* 7 July 1971, p. 24.

CHAPTER 7

1 Denis Willson, then Secretary of Culham lab and a member of the JET site committee, wrote an account of the Common Market maneuverings that preceded JET's birth titled "A European Experiment." I have drawn the details of council meetings from that account.

2 *Times,* 10 April 1984.

CHAPTER 11

1 "Review of the Department of Energy's Inertial Confinement Fusion Program," National Academy of Sciences, March 1986.
2 "Inertial Fusion Fact Sheet," U.S. Department of Energy, 30 September 1988.
3 Max, 1984.
4 Hancox, Redpath, 1985.
5 *Washington Post,* 23 September 1986.

CHAPTER 12

1 Lawrence M. Lidsky, "The Trouble With Fusion," *Technology Review,* September 1983, p. 32.
2 *Washington Post,* 13 November 1983.
3 Correspondence (courtesy Harold Furth).
4 Ansel Adams and Mary S. Alinder, *Ansel Adams: An Autobiography,* Little, Brown, Boston, 1985.
5 Ken Kelley, "The Last Interview," *San Francisco Focus,* June 1984, p. 42.
6 Associated Press, 11 June 1982.
7 Robert L. Hirsch, "Whither Fusion Research? (Speech to the American Nuclear Society, 5 March 1985).

CHAPTER 13

1 *Newsweek,* 5 January 1987.

CHAPTER 14

1 "Palladium: The Fuel of the Future?" *Newsweek,* 8 May 1989.
2 According to a chronology of the episode compiled by Brigham Young University. (Courtesy of Paul C. Richards, director, public communications, Brigham Young University.)
3 *Los Angeles Times,* 3 May 1989.
4 *Science,* 28 April 1989.
5 Parker, private correspondence with author, 11 May 1989.
6 *Time,* 8 May 1989.
7 *Nature,* 27 April 1989.
8 United Press International, 24 May 1989.
9 Edward Teller, private correspondence with author, 3 April 1987.

Glossary

FUSION TERMS*

Alcator: A family of tokamak magnetic confinement devices developed and built at the Massachusetts Institute of Technology and characterized by relatively small diameters and high magnetic fields. The plasmas created in these devices have relatively high current and particle densities.

ampere, kiloampere, megampere: The standard unit for measuring the strength of an electric current representing a flow of one coulomb of electricity per second. A kiloampere = 1,000 amperes. A megampere = 1,000,000 amperes.

Asdex: Axially Symmetric Divertor Experiment (Garching, West Germany). A large tokamak designed for the study of impurities and their control by a magnetic divertor. The H-mode or high mode of operation with neutral beam injection was first observed on Asdex.

atom: The smallest unit of an element that retains the characteristics of that element. At the center of the atom is a nucleus, made up of neutrons and protons, around which the electrons orbit. Atoms of hydrogen, the lightest element, consist of a nucleus of one proton orbited by one electron.

atomic bomb, A-bomb: A weapon with a large explosive power due to the sudden release of energy when the nuclei of heavy atoms such as plutonium-239 or uranium-235 are split. This fission is brought about by the bombardment of the fuel with neutrons, setting off a chain reaction. The bomb releases shock, blast, heat, light, and lethal radiation. The world's first atomic bomb was successfully tested by the United States on July 16, 1945.

* Drawn in part from Glossary of Fusion Energy, U.S. Department of Energy, and Princeton Plasma Physics Laboratory Glossary of Fusion Terms.

blanket: A region surrounding a fusion reactor core within which the fusion neutrons are slowed down, heat is transferred to a primary coolant, and tritium is bred from lithium. In hybrid applications, fertile materials (U-238 or Th-232) are located in the blanket for conversion into fissile fuels.

Bohm diffusion: A rapid loss of plasma across magnetic field lines caused by microinstabilities. Theory formulated by the physicist David Bohm.

breakeven: The point at which the fusion power produced by a reacting plasma equals the input power needed to sustain the plasma's high temperature. To produce net power in a fusion reactor, according to the "Lawson criterion," in a plasma of about 70 million degrees centigrade with a density of 10 to the 14th power particles per cubic centimeter, the plasma must be contained for at least one second.

breeder reactor: A kind of nuclear reactor that produces more fissionable material than it consumes to generate energy. The liquid-metal "fast breeder," a promising type of breeder, splits plutonium-239, producing an intense flow of neutrons and a self-sustaining chain reaction.

centigrade (Celsius): Method of measuring temperature in which the freezing point of water is considered zero degrees and the boiling point of water is 100 degrees. Therefore, zero degrees on the centigrade scale equals 32 degrees Fahrenheit.

chain reaction: A self-sustaining series of chemical or nuclear reactions in which the products of the reaction contribute directly to the propagation of the process. In an atomic bomb, a neutron strikes the nucleus of a uranium atom, splitting it in two with the release of a tremendous amount of heat plus two or more neutrons. These neutrons in turn strike other uranium nuclei, releasing more heat and yet more neutrons, setting up a continuous reaction until the uranium fuel is exhausted.

cold fusion: An unproven method of producing fusion at room temperature by passing an electrical current through two electrodes, one of them made of palladium metal, immersed in a bath of deuterium-saturated water. It is postulated that the deuterium nuclei, driven together in the latticelike structure of the palladium, could come close enough to fuse.

compact torus: Any of a series of fusion experiments involving toroidal magnetic geometries in which the ratio of the radius of the torus to the plasma radius is nearly one.

confinement time: The amount of time the plasma is contained by magnetic fields before its energy leaks away.

controlled thermonuclear fusion: The process in which very light nuclei, heated to a high temperature in a confined region, undergo fusion reactions under controlled conditions, with the associated release of energy which may be harnessed for useful purposes.

D-shaped plasma: A toroidal plasma whose cross section is in the shape of a D.

deuterium atom: An isotope of hydrogen with one proton and one neutron in its nucleus and a single orbiting electron.

diagnostics: Procedures for determining (diagnosing) the state of a plasma during an experiment; also refers to the instruments used for diagnosing.

disruption: A gross instability which gives rise to an abrupt temperature drop and the termination of the plasma.

divertor: Component of a toroidal fusion device that diverts charged particles on the outer edge of the plasma where they become neutralized. This prevents the particles from striking the chamber walls and dislodging secondary particles that would cool the plasma.

DOE: Department of Energy. U.S. cabinet-level department that has overseen atomic energy research since 1977.

Doublet devices: A series of tokamaks designed by GA Technologies (formerly General Atomic) in San Diego making plasmas with noncircular cross sections, including kidney shapes and D-shapes.

EC: European Community. Organization of European countries (formerly the European Economic Community, or EEC) established in 1967 to coordinate policies on the economy, energy, agriculture, and other matters. The original member countries were France, Belgium, West Germany, Italy, Luxembourg, and the Netherlands. Joining later were Denmark, Ireland, the United Kingdom, Greece, Spain, and Portugal.

electron: Elementary particle with a negative electrical charge. Electrons orbit around the positively charged nucleus in an atom.

electron volt (eV): Unit of energy equal to the energy acquired by an electron passing through a potential difference of one volt. The electron volt is also used to express plasma temperatures.

Euratom: European Atomic Energy Community. International organization established in 1958 by members of the European Economic Community to form a common market for the development of peaceful uses of nuclear power.

fission: The division of a heavy atomic nucleus, such as uranium or plutonium, into two approximately equal parts accompanied by the release of a relatively large amount of energy and generally one or more neutrons.

fission bomb: see atomic bomb, A-bomb.

fission reactor: A device that can initiate and control a self-sustaining series of nuclear fissions, converting the energy released into electric power.

fusion: The merging of two light atomic nuclei into a heavier nucleus, with a resultant loss in the combined mass. The fusion is accompanied in general by the release of energy. See also: controlled thermonuclear fusion.

fusion reactor: A proposed type of nuclear reactor in which a self-sustaining series of nuclear fusions would produce energy that could be converted into electric power.

fusion – fission hybrid: A proposed type of nuclear reactor relying on both fusion and fission reactions. A central fusion chamber would produce neutrons to provoke fission in a surrounding blanket of fissionable material. The neutron source could also be used to convert other materials into additional fissile fuels.

greenhouse effect: The warming of the Earth due to the increasing presence of carbon dioxide in the lower atmosphere from the burning of fossil fuels. Heat radiated from the Earth's surface is trapped by the atmosphere, the way heat is held in by the glass windows of a greenhouse.

half-life: The time in which half the atoms of a particular radioactive substance disintegrate to another nuclear form. Half-lives range from millionths of a second to billions of years.

high mode or H-mode: A regime of operation attained during auxiliary heating of divertor tokamak plasmas when the injected power is sufficiently high. A sudden improvement in particle confinement time leads to increased density and temperature, distinguishing this mode from the normal "low mode."

hybrid reactor: see fusion – fission hybrid.

hydrogen: The lightest element, comprised of one proton and one electron. Its isotopes are deuterium, which has one additional neutron and tritium, which has two neutrons.

hydrogen bomb, H-bomb: An extremely powerful type of atomic bomb based on nuclear fusion. The atoms of heavy isotopes of hydrogen (deuterium and tritium) undergo fusion when subjected to the immense heat and pressure generated by the explosion of a nuclear fission unit in the bomb.

IAEA: International Atomic Energy Agency. An autonomous intergovernmental organization established in 1956 with the purpose of advancing peaceful uses of atomic energy, with headquarters in Vienna.

ignition: The point at which the plasma produces so much energy from fusion reactions that it no longer needs any external source of power to maintain its temperature.

impurities: Atoms of unwanted elements in the plasma, inhibiting fusion.

instability: A state of plasma in which any small perturbation amplifies itself to a considerable alteration of the equilibrium of the system, leading to disruptions.

ion: An atom which has become charged as a result of gaining or losing one or more orbiting electrons. A completely ionized atom is one stripped of all its electrons.

ionization: The removing or adding of an electron to a neutral atom, thereby creating an ion.

isotope: One of several versions of the same element, possessing different numbers of neutrons but the same number of protons in their nuclei.

ITER: International Thermonuclear Experimental Reactor. A fusion reactor being planned jointly by the United States, the Soviet Union, Japan, and Europe.

JET: Joint European Torus, a large tokamak in Oxfordshire, England, commonly owned by the European Community.

JT-60: A large Japanese tokamak located north of Tokyo.

laser: Stands for light amplification by stimulated emission of radiation. An optical device that amplifies and concentrates light waves, emitting them in a narrow, intense beam.

laser fusion: A nuclear fusion scheme in which a burst of focused laser light is used to compress and heat a small pellet of fuel.

laser scattering device: Formally known as Thomson scattering device. A diagnostic device used to measure electron temperature in a plasma by directing laser light into the plasma. The laser's photons scatter off the electrons, spreading in a manner proportional to the electron temperature.

limiters: Structures placed in contact with the edge of a confined plasma which are used to define the shape of the outermost magnetic surface.

lithium: A soft, silver-white metal that is the lightest of all metals. It is liquid at 355 degrees Fahrenheit and is used as a breeder of tritium and a coolant in fusion reactor schemes.

low mode, L-mode: The normal behavior of a plasma undergoing ohmic heating, that is, as the plasma's temperature climbs higher, the confinement of the plasma deteriorates.

magnetic bottle: The magnetic field used to confine a plasma in controlled fusion experiments.

magnetic fusion: The use of magnetic fields to contain a plasma that is undergoing fusion.

magnetic mirror devices: Generally, linear fusion devices in which the magnetic fields grow stronger at the end points, causing escaping particles to be reflected back into the main body of plasma.

meltdown: A buildup of heat in the core of a nuclear fission reactor due to an uncontrolled chain reaction of the fission fuel causing the fuel rods to melt down to the reactor floor.

MIT: Massachusetts Institute of Technology.

neutral beam injection: A method for producing neutrally charged atoms of high energy and injecting beams of these atoms into a magnetically confined plasma where they are immediately ionized. The high-energy ions then transfer part of their energy to the plasma particles in repeated collisions, thus increasing the plasma temperature.

Nova: The United States' largest laser fusion facility, located at Lawrence Livermore National Laboratory in California.

nucleus: The central region of an atom containing protons and neutrons.

ohmic heating: Heating resulting from the resistance a medium offers to the flow of electric current. In plasma subjected to ohmic heating, ions are heated almost entirely by transfer of energy from the hotter electrons.

pellet injector: A device that shoots small frozen quantities of hydrogen isotopes at high speed into the inner regions of hot plasma. This method has some penetration advantages over conventional gas injection.

pinch effect: The constriction of a plasma carrying a large current caused by the interaction of that current with its own encircling magnetic field.

plasma: An ionized gas, in this case a gas broken down into an electrically equivalent number of positive ions and free moving electrons. The sun is a plasma.

PLT: Princeton Large Torus. Large toroidal device of the tokamak variety located at the Princeton Plasma Physics Laboratory.

plutonium: A radioactive metallic element whose isotope, plutonium 239, is used in nuclear weapons and as a reactor fuel.

Project Matterhorn: The code name of the United States' first secret controlled fusion project started by Lyman Spitzer at Princeton University in 1951.

proton: An elementary particle found in the nucleus of all atoms. It carries a single positive electrical charge.

radiation: The emission of energy in the form of light, heat, and streams of fast-moving particles from atoms as they undergo internal change. Elements such as uranium and plutonium give off radiant energy in a spontaneous disintegration of their atomic nuclei and are said to be radioactive.

radioactive waste: Equipment and materials from nuclear operations which are radioactive and for which there is no further use.

reactor: see fission reactor.

runaway electrons: Those electrons in a plasma that gain energy from an applied electrical field at a faster rate than they lose it through collision with other particles. These electrons tend to "run away" in energy from the remainder of the plasma.

scaling laws: Laws stating that if two quantities are proportional and are known to be valid at certain orders of magnitude, then they can be used to calculate the value of one of the quantities at another order of magnitude.

scientific feasibility: The successful completion of experiments which reach "breakeven" plasma conditions (minimum values of density, temperature, and plasma confinement time) in laboratory devices which lend themselves to development into net power producing systems. Reactor grade fusion fuels need not necessarily be used in these experiments.

SDI: Strategic Defense Initiative. President Ronald Reagan's plan to launch a space-based missile defense.

stellarator: Device invented by Lyman Spitzer for the containment of a plasma inside a racetrack-shaped tube. The plasma is contained by a magnetic field created by helical windings around the tube.

superconductor: A type of electrical conductor that permits a current to flow with zero resistance.

T-3: A Soviet tokamak located at the Kurchatov Institute in Moscow that first proved viability by producing a plasma temperature of 10 million degrees centigrade.

T-10: A later Soviet tokamak located at the Kurchatov Institute identical to Princeton's PLT but without neutral beams.

T-15: A Soviet tokamak with superconducting magnets currently under construction.

T-20: A large Soviet tokamak that was to have operated under reactor conditions but was abandoned for budget reasons.

temperature, plasma: As applied to a plasma, it is a measure of the random kinetic energy (energy of motion) of the ions or electrons present.

TFTR: Tokamak Fusion Test Reactor. The United States' largest tokamak, located at the Princeton Plasma Physics Laboratory. It is capable of injecting neutral beams into a deuterium-tritium plasma under reactorlike conditions.

thermonuclear fusion: Fusion at high temperature with a significant release of energy.

tokamak: Based on an original Soviet design, a device for containing plasma inside a torus chamber by using the combination of two magnetic fields—one created by electric coils around the torus, the other created by intense electric current in the plasma itself, which also serves to heat the plasma. TFTR and JET are tokamaks.

toroidal system: Name given to the general class of "doughnut-shaped" magnetic devices in which magnetic lines of force close on themselves. Stellarators and tokamaks are examples.

tritium: A radioactive isotope of hydrogen with one proton and two neutrons in its nucleus and one orbiting electron. A more efficient fusion fuel than ordinary hydrogen because of the extra neutrons.

uranium: A radioactive metallic element whose isotope, uranium 235, is a nuclear fission fuel. Plutonium, another fission fuel, can be produced from the more plentiful isotope uranium 238.

ZETA: Zero Energy Thermonuclear Assembly. A British fusion device in which scientists made the erroneous observation in 1958 of thermonuclear fusion reactions.

GLOSSARY OF NAMES

Artsimovich, Lev: Director of the Soviet Union's controlled fusion program from 1950 to 1973.

Bickerton, Roy: Briton who was Deputy Director of the Joint European Torus from 1985 to 1988, previously Director of Experiments.

Bussard, Robert: Fusion physicist and innovator, an assistant director of the U.S. fusion program 1973 – 1974 and founder and chairman of International Nuclear Energy Systems Co. (Inesco).

Clarke, John: Director of the U.S. fusion program since 1982.

Cockroft, John: Director of the United Kingdom's Atomic Energy Research Establishment from 1946 to 1958, overseeing fusion and fission research.

Coensgen, Fred: Plasma physicist at Lawrence Livermore National Laboratory.

Coppi, Bruno: Italian plasma physicist at MIT and inventor of the Alcator tokamak.

Efthimion, Phil: Princeton plasma physicist.

Eubank, Harold: Princeton plasma physicist in charge of neutral beams.

Fermi, Enrico: Nobel Prize-winning physicist and one of the founders of the U.S. atomic fission program.

Fowler, Ken: Director of the magnetic fusion program at Livermore Laboratory from 1970 – 1987.

Furth, Harold: Director of the Princeton Plasma Physics Laboratory since 1981. Previously director of experiments.

Goldston, Rob: Princeton plasma physicist.

Gottlieb, Mel: Director of the Princeton Plasma Physics Laboratory from 1966 to 1981. Succeeded Lyman Spitzer.

Grove, Don: Project manager for the Tokamak Fusion Test Reactor at Princeton from 1982 to 1986.

Guccione, Bob: Publisher of *Penthouse* magazine and *Omni,* large private investor in fusion research.

Hirsch, Robert: Director of the U.S. fusion program from 1972 to 1976.

Iso, Yasuhiko: Director of Japan's Atomic Energy Research Institute.

Kurchatov, Igor: Director of atomic research for the Soviet Union, 1938 – 1960, including development of the atomic bomb.

Lidsky, Lawrence: Nuclear engineer at MIT who challenged the commercial viability of a fusion reactor.

Meade, Dale: Program manager for the Tokamak Fusion Test Reactor at Princeton Plasma Physics Laboratory since 1986; formerly deputy manager under Don Grove.

Muchovatov, Vladimir: Soviet plasma physicist.

Pease, Sebastian: Director of the United Kingdom's Culham Laboratory 1967 – 1987.

Post, Dick: Physicist at the Lawrence Livermore National Laboratory, inventor of the U.S. mirror machine.

Rebut, Paul-Henri: Designer of the Joint European Torus and director of JET since 1985. Succeeded Hans-Otto Wuster.

Robinson, Derek: Plasma physicist at the United Kingdom's Culham Laboratory and member of the team that verified the Russian tokamak breakthrough in 1969.

Rosenbluth, Marshall: Director of the Institute for Fusion Studies at the University of Texas, 1980 – 1987. Nicknamed "The Pope of Plasma Physics."

Sakharov, Andrei: Soviet nuclear physicist who is credited with designing the first tokamak, human rights advocate, winner of the Nobel Peace Prize.

Spitzer, Lyman: Inventor of the stellarator. Founder of the U.S. controlled fusion program and director of the Princeton Plasma Physics Laboratory, 1961 – 1966.

Stodiek, Wolfgang: Princeton plasma physicist.

Storm, Erik: Director of laser fusion research at Livermore Laboratory since 1984, previously deputy director.

Strauss, Lewis: Chairman of the U.S. Atomic Energy Commission, 1953 – 1958.

Tait, Graeme: Princeton plasma physicist.

Tamm, Igor: Soviet physicist. Inventor with Sakharov of the tokamak.

Teller, Edward: Hungarian-born nuclear physicist. Lead the U.S. development of the world's first hydrogen bomb, an early advocate of peaceful fusion research.

Thomson, George Paget: Nobel Prize-winning British physicist who, in 1946, patented the first magnetic fusion device, a circular pinch machine.

Thonemann, Peter: Australian fusion researcher who worked at the United Kingdom's Culham Lab on the earliest British fusion devices, including ZETA.

Velikov, Yevgeny: Soviet plasma physicist, science advisor to President Mikhail Gorbachev, director of Soviet nuclear power and fusion research program.

Ware, Alan: British fusion researcher who worked on original magnetic fusion scheme under George Paget Thomson.

Wuster, Hans-Otto: West German who was the first director of the Joint European Torus, 1978 – 1985.

Yoshikawa, Masaji: Director of the JT-60 Japanese tokamak.

Yoshikawa, Shoichi: Princeton physicist and an original adviser to the Japanese fusion program.

Appendix A: Basic Physics of Fusion

THE FUSION REACTION

The atomic nuclei of light elements can be made to fuse together to form a heavier element while releasing excess energy.

Normally nuclei repel one another because they all carry positive electrical charges. But at high temperature the nuclei can travel very fast, colliding with enough force to overcome their mutual repulsion, and fuse.

In the most readily attainable fusion reaction, nuclei of the hydrogen isotopes deuterium and tritium are brought together at high temperature to fuse and yield helium and energy in the form of energetic neutrons.

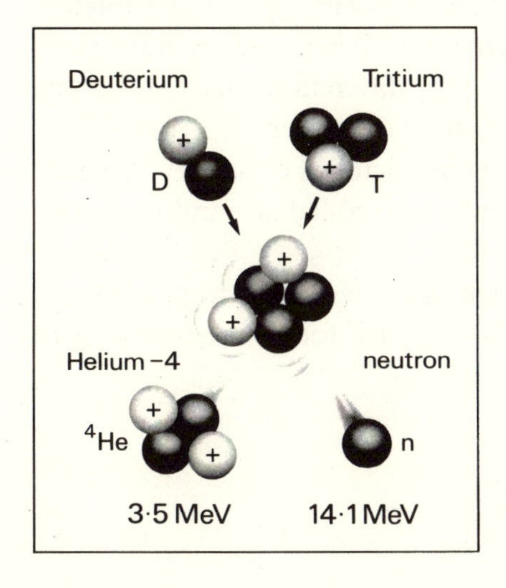

(Diagram courtesy of UKAEA – Culham Laboratory)

Appendix A

REQUIREMENTS FOR A FUSION REACTOR

Occasional fusion reactions, however, are insufficient for a practical fusion reactor, that is, one that creates more energy than it uses to heat the deuterium-tritium fuel. In order to assure abundant fusion reactions, the fuel must be heated to at least 100 million degrees centrigrade, fourteen times hotter than the center of the sun.

Moreover, to maximize the probability of collisions, the high-speed, randomly moving nuclei must be confined closely for a minimum amount of time and at a sufficient density. The required confinement time depends on the density.

At typical 100 million degree conditions, if the density is ten to the fourteenth power particles per cubic centimeter, the fast-moving nuclei must be confined for about one second. This requirement is known as the Lawson criterion. A fuel of denser particles could be confined a shorter amount of time. A thinned-out fuel would have to be held longer.

PLASMA

As a gas is heated to high temperature, the atoms of the gas become "ionized." The electrons, which normally orbit around the atoms' nuclei, are stripped away and a mixture of positively charged nuclear "ions" and negatively charged electrons is formed. This mixture is called a "plasma" and it has properties very different from a gas. Plasma is sometimes referred to as the fourth state of matter.

Plasma, for example, is an excellent conductor of electricity. Because it is made of charged particles, it can be controlled and influenced by magnetic fields.

When the fusion fuel deuterium-tritium is heated to high temperature it forms a plasma which must be confined for a time sufficient to allow abundant fusion reactions to occur.

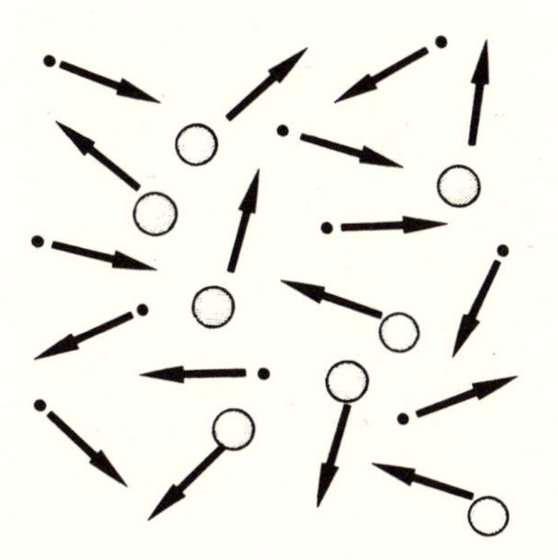

Ionized Plasma

(Diagram courtesy Princeton Plasma Physics Laboratory.)

CONFINING PLASMA

When a magnetic field is applied to a plasma the individual charged particles of the plasma are constrained to move in circles around the lines of magnetic force.

Motion of Charged Particles

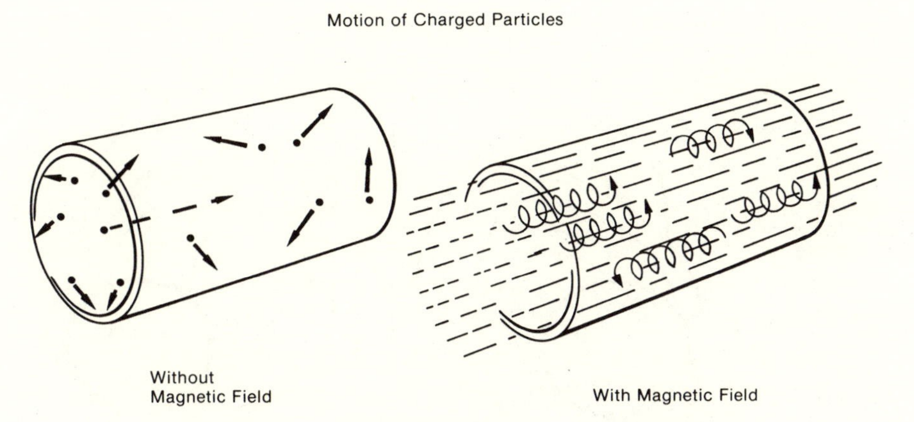

(Diagram courtesy Princeton Plasma Physics Laboratory.)

Appendix B: Fusion Devices

MAGNETIC CONFINEMENT DEVICES

Stellarator

Helical electrical windings around a racetrack-shaped vacuum vessel (early version) create a twisting magnetic field that captures the plasma particles. The stellarator relies exclusively on externally created fields. Later versions were circular and added extra toroidal field coils.

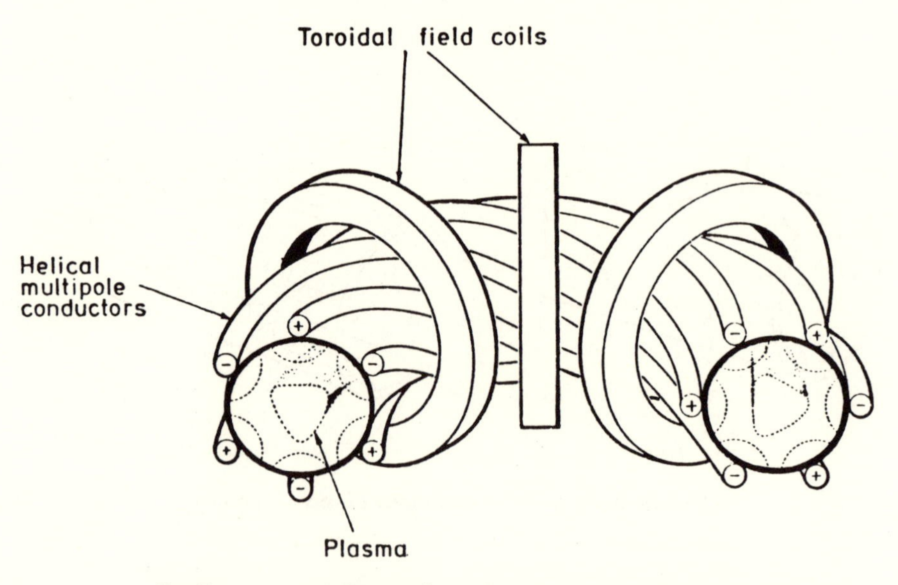

Stellarator configuration showing helical windings ($l = 3$).

(Diagram courtesy UKAEA – Culham Laboratory.)

Magnetic mirror

Usually a linear device. Magnetic field lines are made stronger at the ends to "reflect" the plasma particles back into the device, but there are still losses of particles escaping out the ends.

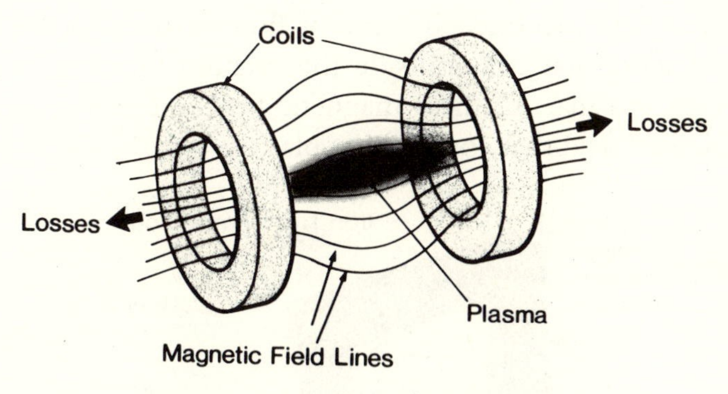

(Diagram courtesy JET Joint Undertaking.)

Toroidal pinch

An electric current carried by the plasma itself produces a corresponding magnetic field that confines the plasma into a column. In a toroidal pinch the vacuum vessel and the plasmas' magnetic field are circular. Additional coils may be added outside the vacuum vessel to create extra stabilizing magnetic fields—making the pinch resemble a tokamak.

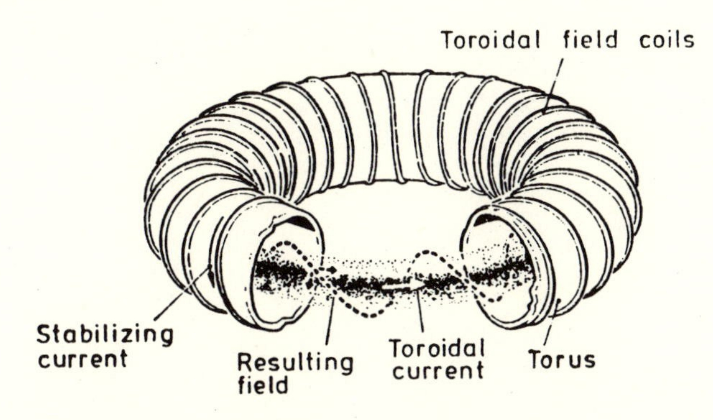

(Diagram courtesy UKAEA – Culham Laboratory.)

Appendix B

Tokamak

The plasma is confined away from the walls of the vacuum vessel by a complex system of magnetic fields. In JET, the main component of the magnetic field, the so-called toroidal field, is provided by 32 D-shaped coils surrounding the vacuum vessel. This field combines with the self-induced field produced by current flowing through the plasma. There are additional coils outside the vacuum vessel to shape and position the plasma. All these magnetic fields interact to produce a twisted, helical field that confines the plasma.

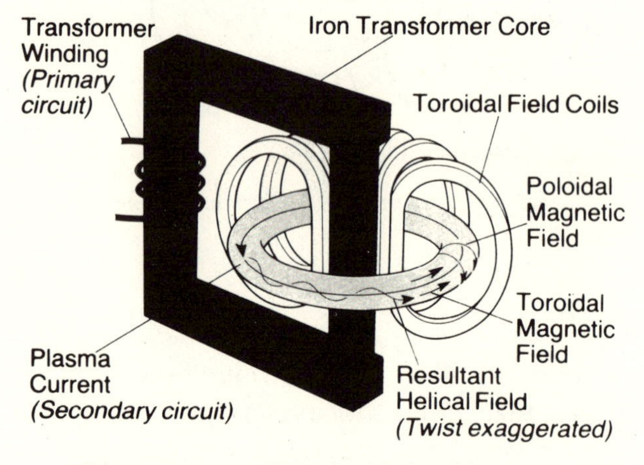

(Diagram courtesy JET Joint Undertaking.)

HOW A FUSION REACTOR WOULD WORK

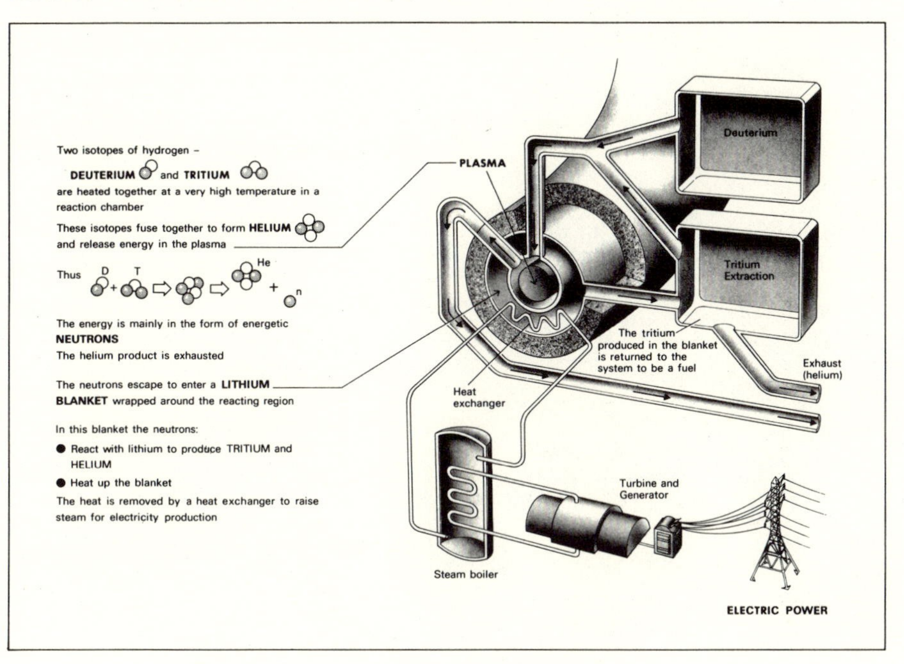

(Diagram courtesy UKAEA – Culham Laboratory.)

LASER FUSION

INERTIAL CONFINEMENT FUSION CONCEPT

Laser energy

Inward transported thermal energy

Atmosphere Formation
Laser or particle beams rapidly heat the surface of the fusion target forming a surrounding plasma envelope.

Compression
Fuel is compressed by rocket-like blowoff of the surface material.

Ignition
With the final driver pulse, the fuel core reaches 1000 – 10,000 times liquid density and ignites at 100,000,000 C.

Burn
Thermonuclear burn spreads rapidly through the compressed fuel, yielding many times the driver input energy.

(Diagram courtesy Lawrence Livermore Laboratory.)

Index

Index

Index